DWDM Fundamentals, Components, and Applications

For a listing of recent titles in the *Artech House Optoelectronics Library*, turn to the back of this book.

DWDM Fundamentals, Components, and Applications

Jean-Pierre Laude

Artech House
Boston • London
www.artechhouse.com

Library of Congress Cataloging-in-Publication Data
Laude, Jean-Pierre.
DWDM fundamentals, components, and applications / Jean-Pierre Laude.
p. cm. — (Artech House optoelectronics library)
Includes bibliographical references and index.
ISBN 1-58053-177-6 (alk. paper)
1. Wavelength division multiplexing. 2. Optical communications.
3. Light—Wavelength. I. Title. II. Series.
TK5103.592.W38 L38 2002
621.382'7—dc21 2001055246

British Library Cataloguing in Publication Data
Laude, J. P. (Jean-Pierre)
DWDM fundamentals, components, and applications. — (Artech House optoelectronics library)
1. Optical communications 2. Multiplexing
I. Title
621.3'827

ISBN 1-58053-177-6

Cover design by Gary Ragaglia

685 Canton Street
Norwood, MA 02062

International Standard Book Number: 1-58053-177-6
Library of Congress Catalog Card Number: 2001055246

10 9 8 7 6 5 4 3 2 1

Contents

Preface

This book is intended for people interested in the future of telephone, data, video, and Internet communication, and to graduate students, scientists, designers, development engineers, and technicians who want to learn more about dense wavelength division multiplexing (DWDM). The readers working in the field of telecommunications, such as components and systems designers, may find useful references along with the text. I hope this book will also be useful to scientists and engineers working in the field of optics, spectroscopy, optronics, electro-optics, and micro/nano technology who would be willing to offer their expertise to solve some of the still-open problems in telecommunications. Conversely, this book may help them to apply to different fields some of the tremendous progress and investments already made in the optical telecommunication field. Perhaps this book will be also useful to marketing managers who want to get information and prospects on the future of optical networks.

It is expected that the book will be mostly useful in understanding components and principles used in existing and/or in next generations of DWDM optical telecommunication networks. I began research on WDM components more than 20 years ago in a company devoted to advanced optical instrumentation. I had the opportunity to participate in the Wavelength Time Division Multiplexing (WTDM) research group within the Research and Development in Advanced Communication Technologies in Europe (RACE) program, and to the Wavelength-agile Optical Transport and Access Network (WOTAN) and Switchless Optical Network for Advanced Transport Architecture Advanced Communications Technologies and

Services (SONATA) groups, within the Advanced Communications Technologies and Services (ACTS) program, in the field of DWDM. This was an ideal platform to participate in the dreams and in the successes of many of the most important European telecommunication groups. As a consultant, I also had the opportunity to participate in the challenge of HighWave Optical Technologies, a new company successfully exploiting the opportunities in the DWDM component and subsystem market.

In 1993, I published my first book entitled *Wavelength Division Multiplexing*. Of course, accelerated research efforts within the international community have produced many new concepts and devices in the field. This book could not be a revised edition of my first work. I had to analyze and summarize many new developments, keeping only when necessary some of the basic material of the first work.

Acknowledgments

I am grateful to Christine, my wife, for her love. Our children, Jean-Christophe, Vincent, Thomas, Blandine, Marjolaine, and Grégoire, have been a great source of support and encouragement. They could understand why this book was so impossibly demanding, as most of them are involved in science and technology. Dr. Vincent Laude deserves a special mention as he played an important role in the critical reading and correction of the entire book. I would like to express my gratitude for his patience and readiness to help.

1

Introduction

Si possible je voudrais montrer des futurs désirables qui n'ont pas encore été imaginés à seule fin d'ajouter de nouveaux buts. (If possible, I want to show desirable prospects that have never been conceived for the sole purpose of adding new aims.)

Tom Shannon
Spring 2000

The long-awaited deployment of dense wavelength division multiplexing (DWDM) in long-haul networks is fast becoming a reality. In the access network DWDM becomes a serious option to accommodate increasing multimedia, broadband, and IP traffic.

Fueled by the need for rapidly increasing volumes of information transfer, and by new requirements that are far less predictable than they have been for telephony, the development of optical fiber communication moves ever forward. DWDM that corresponds to the superimposition of optical signals at different wavelengths closely spaced at 100 GHz or less on each fiber, becomes, as a matter of fact, the unavoidable solution, not only for a capacity increase at lower cost, but also for switching and routing in the optical domain.

The optical add/drop multiplexer (OADM) and optical cross-connect technology are seen as a viable option in the large network optical nodes, when the digital cross-connectors are unable to scale to higher port-density and higher speed interface requirements at affordable costs, or in general, when the optical/electrical/optical conversion would be unnecessary and questionable.

Today's telecommunications networks have benefited from research in symbiosis in the fields of materials, electromechanics, microelectronics, computing, and optics.

Nowadays, low nonlinearity fibers, dispersion management on fibers, lasers with high wavelength stability over years, new tunable light sources, new erbium-doped amplifiers, distributed Raman amplifiers, semiconductor amplifiers and gates, wavelength converters, couplers, space-switches, and wavelength division multiplexers offer new solutions. DWDM with more than 160 channels, OADM with more than 32 channels, and wavelength-selective cross-connects (WSXC) with 32 × 32 input-to-output ports are currently in commercial use. Wavelength-interchanging cross-connects (WIXC) will become available soon, as the first industrial tunable lasers and wavelength converters are already proposed on the market.

We did our best to organize the text of this book into a logical sequence of chapters. After the short introduction of this chapter, the basic principles and background in wavelength division multiplexing are reviewed in Chapter 2. Chapter 2 also includes an introduction to soliton multiplexing, one of the options for long-distance transmissions. In Chapter 3, we review the current choices available in DWDM passive components. In Chapter 4 we present the main active sources including semiconductors lasers, glass-doped lasers, broadband sources for spectral slicing, and the wavelength conversion technology. In Chapter 5 we review the different solutions in optical amplification. Optical switches, routers, cross-connects, and add/drops are described in Chapter 6, after discussions on topological configurations and architectures of optical networks in which they are utilized.

In Chapter 7, the optical fiber nonlinearities limiting the performance of DWDM systems are presented. Chapter 8 gives examples of applications of the DWDM technology and results in long-haul transmission networks and in different passive or wavelength-switched voice and data networks including the Internet.

2

Basic Principles and Background

2.1 Wavelength Division Multiplexing: Basic Principles

Telecommunications makes wide use of optical techniques in which the carrier wave belongs to the classical optical domain. The wave modulation allows transmission of analog or digital signals up to a few gigahertz or gigabits per second on a carrier of very high frequency, typically 186 to 196 THz. In fact, the bit rate can be increased further, using several carrier waves that are propagating without significant interaction on the same fiber. It is obvious that each frequency corresponds to a different wavelength. This technique is called frequency division multiplexing (FDM) or wavelength division multiplexing (WDM). The latter term is currently preferred in most cases. DWDM is reserved for very close frequency spacing (typically less than 100 GHz corresponding to 0.8 nm at wavelengths near 1.5 μm). The term "frequency division multiplexing" is used in a few cases, such as multiplexing with optical frequency shift keying and coherent detection. But the terminology is not completely stabilized.

With WDM, it is possible to couple sources emitting at different wavelengths—λ_1, λ_2, λ_j, ... λ_n—into the same optical fiber. After transmission on the fiber, the λ_1, λ_2, ... λ_n signals can be separated towards different detectors at the fiber extremity (Figure 2.1). The component at the entrance must inject the signals coming from the different sources into the fiber with minimum losses: This is the multiplexer. The component separating the wavelengths is the demultiplexer. A simple optical coupler may replace the multiplexer, but losses will increase. Obviously, when the light propagation

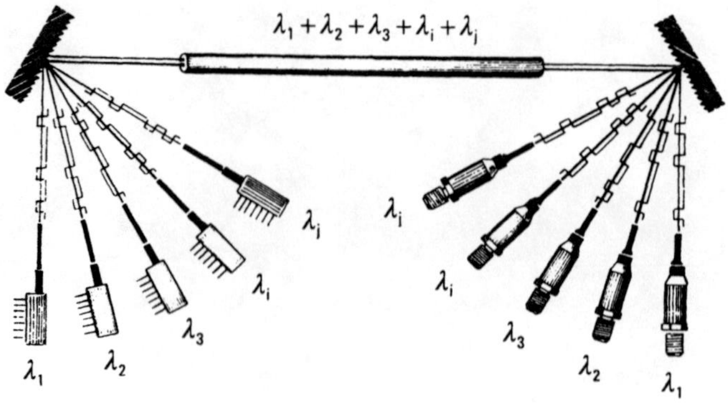

Figure 2.1 WDM.

is reversed, the multiplexer becomes the demultiplexer, and the reverse is also true. It is important to note, however, that the coupling efficiency is not necessarily preserved in reverse operation. For example, if the multiplexer uses single-mode (SM) entrance fibers and a multimode output fiber, the coupling losses would be excessive in the reversed usage. Multiplexers designed with identical input and output fibers are usually reversible. Simultaneous multiplexing of input channels and demultiplexing of output channels can be performed by the same component, the multi/demultiplexer.

2.2 History of WDM in a Few Words

The optical multiplexing concept is not new. To our knowledge, it dates back to at least 1958 [1, 2]. Perhaps we can say that the idea of sending multiple signals, as shown in Figure 2.1, was straightforward, as it was a transposition of techniques used in classical telecommunications with electronics signals. But the technical problems to be solved were very difficult, and it took experts a great deal of time to solve them. About 20 years later, the first practical components for multiplexing were proposed primarily in the United States, Japan, and Europe. In 1977, the first grating-WDM passive component was developed by Tomlinson and Aumiller [3].

2.3 WDM and Time Division Multiplexing

Is it easier to multiplex the signal in the electronic domain—time division multiplexing (TDM), or in the optical domain (FDM or WDM)? The answer to this question is not easy and the optimum solution is generally found in the association of the different techniques.

For low bit-rate services (< 2 Mbps), it is generally better to use only TDM techniques. For uncompressed, high-definition television (HDTV) broadcasting, WDM is highly recommended. The video compression techniques minimize the bandwidth requirement. However, at the time of this writing, CATV and HDTV still require 4 Mbps and 25 Mbps, respectively. Applications such as video networks linking workstations, television studio center signal routing systems, video conference networks, interactive video training systems, bank information service networks, and data-transfer networks between computers, integrated service digital networks (ISDN), teledistribution, and generally all broadband networks increasingly use both time and wavelength multiplexed optical lines. Today, the predicted demand per subscriber in 2010 is on the order of 100 Mbps. That will not be possible without the deployment of DWDM optical fiber networks.

It is understood that a practical network is very often made up of an association of architectures that constitute the physical medium of the network between stations. The topology is called "virtual" when it is concerned only with logical connections between stations. One example of an optical-multiplexing application is to create virtual topologies on request. The network configuration can be modified independently of its physical topology by changing the emitted or received optical frequencies. In these architectures WDM cross-connectors, WDM routers, and WDM add/drops become more and more important.

2.4 Wavelength Domain and Separation Between Channels

With modern, commercially available telecommunication fibers, it is possible to transmit information over a large spectral range (Figure 2.2) with two domains with low attenuation, one around 1.3 μm and another around 1.55 μm. Between these two domains there is generally a high attenuation at 1.39 μm due to the residual OH radical in the fiber. SM silica fibers with minimum loss of about 0.16 dB/km at 1.55 μm are available. Losses < 0.4 dB/Km at 1.5 μm and < 0.5 dB/Km at 1.31 μm are specified in the ITU-T G.652 recommendations. In medium- and long-distance DWDM transmissions, SM fibers are generally used in the domain 1,520 to 1,620 nm (see paragraph 5), due to the availability of efficient optical amplifiers and sources.

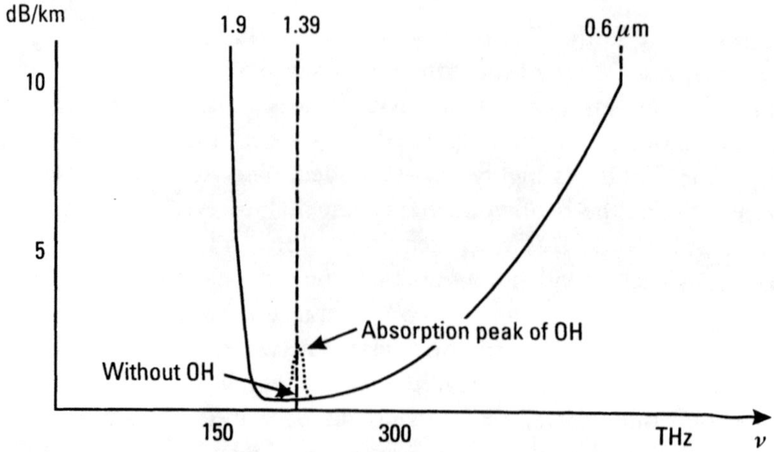

Figure 2.2 Typical loss of a low OH content fiber.

For applications such as metropolitan area networks, special SM fibers with very low OH content are commercialized. They can be used from 1,335 to 1,625 nm (for instance, AllWave fiber from Lucent Technologies).

On multimode silica vapor phase axial deposition (VAD) fiber without OH impurity, losses lower than 4 dB on 2.4 km were obtained from 0.65 to 1.9 μm as long ago as 1982 [4].

On multimode fibers, a graded index design allows a temporal dispersion minimization, but the optimum profile depends greatly on wavelength and material, and the dispersion varies with wavelength.

The separation between channels is now 0.8 nm or more on most of the installed networks using WDM. International Telecommunication Union (ITU) standardization proposes a frequency grid with separations of 100 GHz (about 0.8 nm) with multiples and submultiples. Now the published minimum channel spacing is about 0.1 nm. However, at the beginning of the twenty-first century, nothing lower than 0.2-nm (25 GHz) spacing was commercially available. At first glance, a fiber without OH would allow 1,000 channels at 50-GHz spacing to be multiplexed over its large spectral range!

Of course there are some limitations of WDM (Figure 2.3). The main problem is crosstalk (parasitic light) coming from technical defects in the demultiplexers, but also from physical problems such as wavelength conversion along the transmission fiber by four-wave mixing, Brillouin or Raman effect, or other nonlinear effects. But the theoretical minimum channel spacing is at last related to "uncertainty" relationships.

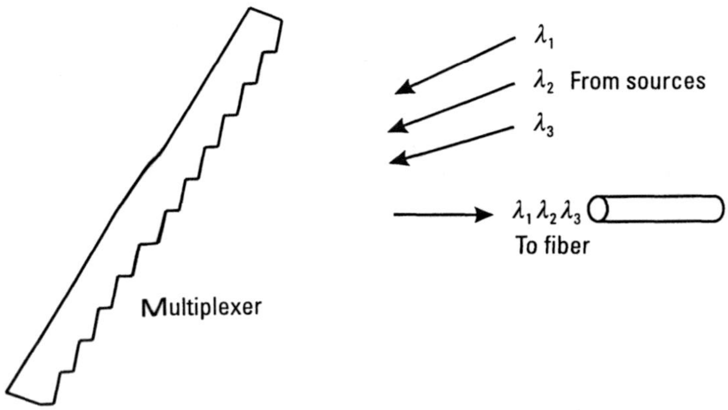

Figure 2.3 Principle of multiplexing by diffraction on an optical grating: Wavelengths λ_1, λ_2, λ_3 coming from different directions are diffracted in the same direction into a single-transmission fiber.

2.5 Wavelength Allocation

Thus far, the ITU-T standards recommends 81 channels in the C band with a constant spacing of 50 GHz anchored at 193.1 THz. This range can be extended to the L band (191.4 to 185.9 THz) where sources and amplifiers become available now. This will add 111 channels at 50-GHz spacing (see Table 2.1).

2.6 Optical Wavelength/Optical Frequency Conversion

Depending on one's background (the classical optical field or the microwave field), one generally prefers to represent light vibrations according to their wavelengths in vacuum (the wavelength varies with the medium) or according to their frequencies (invariant of the medium). In order to evaluate the different results given in scientific papers, it is useful to be able to translate quickly.

It is well known that:

$$\lambda = \frac{c}{\nu}$$

where c is the speed of light ($c = 2.9972458\ 10^6$ m/s), λ is the wavelength in vacuum, and ν is the optical frequency.

Table 2.1
Frequency Standards with Corresponding Wavelengths

C Band	**(196.1–192.1 THz)**	**L Band**	**(191.4–185.9 THz)**
Frequency (THz)	Wavelength (Vacuum) (nm)	Frequency (THz)	Wavelength (Vacuum) (nm)
196.1	1,528.77	191.4	1,566.31
196	1,529.55	191.3	1,567.13
195.9	1,530.33	191.2	1,567.95
195.8	1,531.12	191.1	1,568.77
195.7	1,531.9	191	1,569.59
195.6	1,532.68	190.9	1,570.42
195.5	1,533.47	190.8	1,571.24
195.4	1,534.25	190.7	1,572.06
195.3	1,535.04	190.6	1,572.89
195.2	1,535.82	190.5	1,573.71
195.1	1,536.61	190.4	1,574.54
195	1,537.4	190.3	1,575.37
194.9	1,538.19	190.2	1,576.2
194.8	1,538.98	190.1	1,577.03
194.7	1,539.77	190	1,577.86
194.6	1,540.56	189.9	1,578.69
194.5	1,541.35	189.8	1,579.52
194.4	1,542.14	189.7	1,580.35
194.3	1,542.94	189.6	1,581.18
194.2	1,543.73	189.5	1,582.02
194.1	1,544.53	189.4	1,582.85
194	1,545.32	189.3	1,583.69
193.9	1,546.12	189.2	1,584.53
193.8	1,546.92	189.1	1,585.36
193.7	1,547.72	189	1,586.2
193.6	1,548.51	188.9	1,587.04
193.5	1,549.32	188.8	1,587.88
193.4	1,550.12	188.7	1,588.73

Table 2.1 (continued)

C Band	**(196.1–192.1 THz)**	**L Band**	**(191.4–185.9 THz)**
Frequency (THz)	Wavelength (Vacuum) (nm)	Frequency (THz)	Wavelength (Vacuum) (nm)
193.3	1,550.92	188.6	1,589.57
193.2	1,551.72	188.5	1,590.41
193.1	1,552.52	188.4	1,591.26
193	1,553.33	188.3	1,592.1
192.9	1,554.13	188.2	1,592.95
192.8	1,554.94	188.1	1,593.79
192.7	1,555.75	188	1,594.64
192.6	1,556.55	187.9	1,595.49
192.5	1,557.36	187.8	1,596.34
192.4	1,558.17	187.7	1,597.19
192.3	1,558.98	187.6	1,598.04
192.2	1,559.79	187.5	1,598.89
192.1	1,560.61	187.4	1,599.75
—	—	187.3	1,600.6
—	—	187.2	1,601.46
—	—	187.1	1,602.31
—	—	187	1,603.17
—	—	186.9	1,604.03
—	—	186.8	1,604.88
—	—	186.7	1,605.74
—	—	186.6	1,606.6
—	—	186.5	1,607.47
—	—	186.4	1,608.33
—	—	186.3	1,609.19
—	—	186.2	1,610.06
—	—	186.1	1,610.92
—	—	186	1,611.79
—	—	185.9	1,612.65

Therefore,

$$\Delta\lambda = \frac{\lambda^2}{c}\Delta\nu$$

In practice,

$$\lambda = \frac{299{,}792.458}{\nu}$$

with λ in nm and ν in GHz

$$\Delta\lambda_{\text{nm}} = \frac{\Delta\nu_{\text{GHz}}\lambda^2_{\mu\text{m}}}{0.3}10^{-3}$$

Thus, 100-GHz spacing at 1.55-μm wavelength is equivalent to a spacing of 0.8 nm.

2.7 How Many Channels?

There is actually about 15,000 GHz of optical frequency bandwidth in each 1,300- and 1,550-nm window. With a 10 Gbps bit rate, the uncertainty relationship gives approximately 10 GHz as the limit for optical frequency spacing. This would mean 1,500 channels! But the fiber nonlinearity has always set the limit to a few hundred channels in practical applications.

For the component itself at this limit, the optical crosstalk is the main problem. Acceptable 160-channel grating-WDM components are already manufactured. We believe that many more channels are feasible. Three thousand-channel classical-grating spectrometers are commonly used, why not WDM networks with a few hundred channels? In fact, the main problem is acquiring enough stable fixed or tunable sources. Today, the practical limit is a few tens of sources spaced at 50 GHz. The number of channels will depend on the progress on the sources. It is worth pointing out that an optical frequency chain-generation from a single supercontinuum source with over 1,000 channels at 12.5-GHz spacing has been proposed by H. Takara, et al. in *ECOC'2000* (see Chapter 4).

2.8 Some Definitions

2.8.1 Losses

The multiplexer must combine the signals with minimal losses. Those losses P_j are expressed in decibels (dB) at each wavelength λ_j by:

$$P_j = 10\log\left(\frac{\Phi_j}{\Phi_0}\right)$$

where Φ_j is the optical power injected into the transmission line and Φ_0 is the incident power at λ_j.

2.8.2 Crosstalk

At the other end of the fiber, the signals at the different wavelengths are separated by a demultiplexer which, like the multiplexer, must have minimal losses. The optical crosstalk D_{ij} of a channel i on a channel j is:

$$D_{ij} = 10\log\left(\frac{\Phi_{ij}}{\Phi_{jj}}\right)$$

where Φ_{ij} is the residual optical power of channel i at wavelength λ_i in channel j and Φ_{jj} the exit optical power in channel j at wavelength λ_j.

The total optical crosstalk in channel j is:

$$D_j = 10\log\left(\frac{\sum_{i\neq j}\Phi_{ij}}{\Phi_{jj}}\right)$$

This defect is merely due to the demultiplexer when sources with spectral widths much smaller than the multiplexer spectral passbands are used. But it also becomes necessary to take into account the multiplexer crosstalk in other cases. In such cases (such as LED slicing) the optical crosstalk is a complicated function of sources, multiplexer and demultiplexer.

The electrical crosstalk also depends on the receivers. At equivalent receiver sensitivity, in general, electrical crosstalk is twice as small in decibels as the optical crosstalk. The electrical crosstalk of a system also depends on the relative power and spectral width of the emitters, on the fiber spectral transmission, and on the receiver's sensitivity variation with wavelength. For example, the difference is about 50 dB in a silicon receiver between 0.8 μm (high sensitivity) and 1.3 or 1.5 μm (almost no sensitivity).

When the same optical component performs multiplexing of some wavelengths and demultiplexing of the other wavelengths, we use a similar crosstalk definition, but the term "near-end crosstalk" is used for the parasitic effect of sources on receivers in the vicinity of these sources. The crosstalk coming from sources located at the other end of the line is called "far-end crosstalk." It is obvious that the intrinsic near-end crosstalk of a component must be several orders of magnitude lower than its far-end crosstalk specification because the noise of sources not attenuated by the transmission line is superimposed through the component crosstalk to the signal of a source attenuated in the transmission line. Components with intrinsic near-end optical crosstalk lower than −120 dB have been manufactured [5]. The near-end crosstalk also depends on the different reflections along the line, in particular on the number of connectors and on their location along the transmission line.

2.9 Solitons

2.9.1 Soliton Propagation

The pulse broadening induced by chromatic dispersion can be compensated for by a self-phase shift induced by refraction index variations upon the field intensity (known as the Kerr effect) for particular pulses that have a critical power and shape called solitons. With silica glass the nonlinear index is $n \approx n_0 + 3.2 \times 10^{-16} I$, with I in watt/cm^2 and in which n_0 is the index at an arbitrary low-intensity I. The response time of this phenomenon is arising from electronic structure distortion in less than 1 fs (10^{-15}s = 1 fs) and from nuclei motion in about 100 fs in silica.

In soliton propagation, the velocity of the trailing half of the pulse tends to be increased and the velocity of the leading half of the pulse tends to be decreased by the nonlinear effect; this compensates for the reverse effect due to fiber dispersion. Soliton propagation implies a particular power and pulse shape. With pulses of a few picoseconds' duration, the index variation is relatively fast compared to the pulse duration. A given power level is required to reach the soliton propagation condition. The fundamental soliton of peak power P propagates along a lossless fiber without shape change for an arbitrary long distance. At $N^2 \times$ P peak power, where N is an integer, we obtain higher order (N) soliton propagation. The higher order solitons have a complex behavior: They undergo periodic shape variations along the fiber length but return to their original shape after a length Z_{so} called the soliton period. With pulse width T_0:

$$Z_{so} = \frac{\pi}{2} \frac{T_0^2}{|\beta_2|}$$

where

$$\beta_2 = \frac{\lambda^3}{2\pi c^2} \frac{d^2 n}{d\lambda^2}$$

is the group velocity dispersion [6].

The first theoretical study of soliton propagation in fiber was carried out by A. Hasegawa and F. Tappert in 1973 [7]. However, an amazing phenomenon involving a wave which propagated without shape modification was reported by Russel, a Scottish naval architect in 1834. He had observed a slow-moving wave that propagated about two miles down a canal without losing its shape. He called it a solitary wave. Since then, many researchers have contributed to the study of such propagation. The contribution of Bell Labs to this research has been important; see, for instance, [8, 9]. The earliest research pointed out the soliton robustness to polarization-mode dispersion. This was confirmed by long-term measurements [10].

2.9.2 WDM of Solitons

Coupled with WDM, this propagation mode is very promising for ultra-long-distance transmission; a transmission of 1 million kilometers was demonstrated in 1992 [11]. The same year, NTT demonstrated 20-Gbps soliton transmission over 1,020 km, using an erbium-doped amplifier and forecast Tbps transmission with WDM, and a 70-Gbps fiber-based source of fundamental solitons at 1,550 nm was designed by scientists from Russia and England [12]. Today, much progress has been made. Terrestrial DWDM transmissions with 1.6 Tbps capacity at 50-GHz channel spacing in 8 THz ≈ 64 nm that could be error-free over at least 5,000 km are being tested [13]. DWDM systems with few Tbps are becoming commercially available (for example, from Algety S. A., a French division of Corvis, Inc.).

Ultra-long-distance WDM solitons using broadband fiber amplifiers were studied in several laboratories (an early work is reported in [14]). It was proposed almost 20 years ago to use the flux of nondispersive solitons, [15, 16], with loss compensated by the Raman gain obtained by periodic injection of continuous wave (CW) pump power to obtain an ultra-long-distance high bit rate for all optical transmissions (50 ps solitons over more than 6,000 km in 1989 [17]). This technique is now widely used. In 1992, error-free soliton transmission over more than 11,000 km at 10 Gbps in WDM at 1,555.32 and 1,555.68 nm was demonstrated [18]. It was shown, [19, 20], that with uniformly canceled loss, solitons with different wavelengths emerge from mutual collisions without modification. The collision length L_c is defined as the distance the solitons of different channels traveling in the fiber overtake and pass through each other (the overlap is defined to begin and end at the half power point). With ordinary solitons: $L_c = 2\tau / D\,\Delta\lambda$ where $\Delta\lambda$ is the distance between the wavelengths of two colliding solitons, τ is the pulse full widths at half-maximum (FWHM), and D the fiber dispersion [21, 22]. With 0.8-nm spacing, $D = 0.5$ ps/nm.km, $\tau = 18$ ps this gives $L_{coll} = 90$ km. In lossless and unperturbed conditions, the only result after collision is a time displacement δt between solitons of different frequencies ν and $\nu + \delta\nu$ such that $\delta t = \pm 0.1786\ 1/\tau\ (\Delta\nu)^2$, where τ is the FWHM pulse intensity. The corresponding maximum frequency shift during the collision is $\delta\nu = \pm 0.105/\tau^2\ \Delta\nu$ [22]. On the other hand, in the case of variations in loss and dispersion, a net frequency shift and associated time displacement result from the soliton collision. A perturbation length L_p being defined as the amplification period itself or being applied to other perturbations such as dispersion, and a collision length L_c corresponding to the overlap segment limited at half-power, Mollenauer, et al. showed that if $L_c \geq 2\ L_p$, the collision-induced permanent frequency shifts of the solitons leading to large timing shifts at the end of the transmission link are negligible.

From $L_c = 2\tau / D\,\Delta\lambda$, this sets a maximum allowable $\Delta\lambda$ such as: $\Delta\lambda = \tau / D\,L_p$.

In the practical case of 9,000 km, $L_p = 40$ km, dispersion = 1 ps/(nm · km), $\tau = 50$ ps at 4 Gbps at each frequency, $L_c = 930$ km, $\lambda \cong 1.55\ \mu$m, $\Delta\nu = 146$ GHz (1.17 nm) [21]. If $\Delta t_{max} = \pm 7.5$ ps is set, this corresponds to a spacing between channels limited to $0.3 < \Delta\lambda < 1.2$ nm. However, a frequency-dependent gain obtained, for instance, with dispersion-shifted distributed Er^{+3}-doped fibers may be used to compensate for the soliton frequency shift [23].

From another point of view, limitations in the transmission capacity of soliton-based communications using both dispersion-shifted fibers (DSF) and normal fibers are examined in [24]. This study is not specific to WDM systems. In the best case, corresponding to DSF fibers, loss limitation and collision limitation on distance are, respectively, 50 and 24 km at 50 Gbps. It was shown that "the main advantage of soliton transmission is not that it removes the dispersion effects, but rather a technique to combat the ills of self- and cross-phase modulation (XPM) that prevents multichannel operation in linear systems" [24]. However, with the techniques of dispersion management available now, perhaps this needs to be reexamined.

In soliton transmission it is necessary to keep enough time separation between successive solitons (typically five to six times the pulse width). For channel capacity, this is a relatively inefficient use of time. In order to enhance the transmission efficiency, it was proposed to add polarization or amplitude multiplexing to WDM [25].

2.9.3 Frequency-Guiding Filtering

The use of optical filters periodically spaced along the transmission line (typically one for each amplifier) allows for a significant reduction in both amplitude and time jitter [26, 27].

In the sliding frequency filtering, a gradual shift of the wavelength is applied on successive filters along the line. The solitons automatically follow the small wavelength shift. This filtering also reduces the noise growth. With wavelength periodic filters such as low-finesse Fabry-Perot (FP) filters the filtering is compatible with WDM transmission as it can be applied simultaneously to all channels. The frequency, time, and amplitude jitter is reduced on all channels.

2.9.4 Dispersion-Managed Solitons

In this technique, the absolute value of the dispersion D is made locally high, periodically positive and negative along the path in order to get a low-average value of D [28, 29]. This provides an efficient suppression of four-wave mixing (FWM), see the chapter on nonlinearity) with a Gordon-Haus jitter acceptably small. In WDM the average value of D must remain nearly constant in the wavelength range. Average values of D with zero slope (average $\delta D/\delta\lambda = 0$) are necessary. This can be obtained with the use of one standard fiber and two special fibers along each span. The pulse width (maximized locally when the local value of D is increased) needs to be small enough to avoid adjacent-pulse interaction. Dispersion-slope compensation was proven to be necessary for high-capacity WDM systems (80 to 160 Gbps) [30]. Dispersion management allows single-channel 40-Gbps transmission and multichannel Tbps submarine and long-haul terrestrial transmissions (see, for instance [31–33]).

2.9.5 Conclusion

The interest in soliton transmission comes in part from its compatibility with WDM transmission, and its important feature of being resilient, to some extent, to polarization-mode dispersion and to self- and cross-phase modulation. Filtering and dispersion-management techniques open the way to very-high capacities over transoceanic distances.

References

[1] Denton, R. T., and T. S. Kinsel, "Optical Multiplexing and Demultiplexing," *Proc. IEEE,* Vol. 56, 1958, p. 146.

[2] De Lange, O. E., "Wideband Optical Communication Systems: Part 2, Frequency Division Multiplexing," *Proc. IEEE,* Vol. 58, 1970, p. 1683.

[3] Tomlinson, W., and G. Aumiller, "Optical Multiplexer for Multimode Fiber Transmission Systems," *Appl. Phys. Lett.*, Vol. 31, 1977, p. 169.

[4] Chida, K., F. Hanawa, and M. Nakahara, "Fabrication of OH-Free Multimode Fiber by Vapor Phase Axial Deposition," *IEEE J. Quantum Electronics, Vol QE-18, No. 11, Nov. 1982, pp. 1883–1899.*

[5] Gacoin, P., et al., "Un Multiplexeur Trois Voies à Faibles Pertes," *Revue Physique Appliquée,* Vol. 19, 1984, pp. 99–109.

[6] Dianov, E. M., "Optical Solitons in Fibres," *Europhysics News,* Vol. 23, 1992, pp. 2326.

[7] Hasegawa, A., and T. Tappert, "Transmission of Stationary Nonlinear Optical Pulses in Dispersive Dielectric Fibers. I. Anomalous Dispersion," *Appl. Phys. Lett.*, Vol. 23, No. 3, 1973, pp.142–149.

[8] Stolen, R. H., and C. Lin, "Self-Phase-Modulation in Silica Optical Fibers," *Phys. Review,* Vol. A17, No. 4, 1978, pp. 1448–1453.

[9] Mollenauer, L. F., R. H. Stolen, and J. P. Gordon, "Experimental Observation of Picosecond Pulse Narrowing and Solitons in Optical Fibers," *Phys. Rev. Lett.*, Vol. 45, No. 13, 1980, pp. 1095–1098.

[10] Sunnerud, H., et al., "Experimental Quantification of Soliton Robustness to Polarization-Mode Dispersion," *ECOC'2000 Proc.*, Vol. 3, Munich, 3–5 Sept. 2000.

[11] Miller, G., "Fiberoptics Industry Report," *Laser Focus World*, Pennwell Pub., July 1992, p. 44.

[12] Chernikov, S. V., et al., "70 Gbit/s Fibre Based Source of Fundamental Solitons at 1550 nm," *Elect. Lett. (U.K.)*, Vol. 8, No. 13, June 1992, pp. 1210–1212.

[13] Mollenauer, L. F., "Dispersion Managed Solitons for Ultra Long Distance Terabit WDM," *ECOC'2000 Proc.*, Tutorial 5, Munich, Sept. 3–7, 2000.

[14] Moores, J. D., "Ultra-Long Distance WDM Soliton Transmission Using Inhomogeneously Broadened Fiber Amplifiers," *IEEE J. Lightwave Technol. (U.S.)*, Vol. 10, No. 4, 1992, pp. 482–487.

[15] Hasegawa, A., "Amplification and Reshaping of Optical Solitons in a Glass Fiber - IV: Use of Stimulated Raman Process," *Opt. Lett.*, Vol. 8, 1983, p. 650.

[16] Mollenauer, L. F., J. P. Gordon, and M. N. Islam, "Soliton Propagation in Long Fibers with Periodically Compensated Loss," *IEEE J. Quantum Electronics*, Vol. QE-22, 1986, pp. 157–173.

[17] Mollenauer, L. F., and K. Smith, "Soliton Transmission over More Than 6000 Km in Fiber with Loss Periodically Compensated by Raman Gain," *ECOC'89 Proc.*, Vol. 2, 1989, pp. 7178.

[18] Mollenauer, L. F., E. Lichtman, and G. T. Harvey, "Demonstration of Error-Free Soliton Transmission over More Than 15,000 Km at Gbit/s Single/Channel and over More Than 11,000 Km at 10 Gbit/s in Two-Channel WDM," *Elec. Lett. (U.K.)*, Vol. 28, No. 8, April 9, 1992, pp. 792–794.

[19] Mollenauer, L. F., J. P. Gordon, and M. N. Islam, "Soliton Propagation in Long Fibers with Periodically Compensated Loss," *IEEE J. Quant. Elect.*, Vol. QE-22, 1986, pp. 157–173.

[20] Zahkarov, V. E., and A. B. Shabat, "Exact Theory of Two Dimensional Self Focusing and One Dimensional Self Modulation of Waves in Nonlinear Media," *Zh. Ehsp Teor.* Fiz., Vol. 61, pp. 118–134, July 1971—*Sov. Phys. JETP*, Vol. 34, January 1972, pp. 62–69.

[21] Mollenauer, L. F., S. G. Evangelides, and J. P. Gordon, "Wavelength Division Multiplexing with Solitons in Ultra Long Distance Transmission Using Lumped Amplifiers," *IEEE J. Lightwave Technol. (U.S.)*, Vol. 9, No. 3, March 1991, pp. 362–367.

[22] Mollenauer, L. F., and P. V. Mamyshev, "Massive Wavelength-Division Multiplexing with Solitons," *IEEE J. Quantum Electronics*, Vol. 34, No. 11, Nov. 1998.

[23] Ding, M., and K. Kukuchi, "Analysis of Soliton Transmission in Optical Fibers with the Soliton Self-Frequency Shift Being Compensated by Distributed Frequency Dependent Gain," *IEEE Photon. Technol. Lett. (U.S.)*, Vol. 4, No.5, May 1992, pp. 497–500.

[24] Olsson, N. A., and P. A. Andrekson, "Prospect for High Bit-Rate Soliton Communication," *ECOC'92 Proc.*, Vol. 2, 1992, pp. 746–750.

[25] Hasegawa, A., "Multiplexing in Soliton Transmission Systems," *ECOC'95 Proc.*, Vol. 1, Brussels, Sept. 17–21 1995, pp. 55–58.

[26] Mecozzi, A., and H. A. Haus, "Effect of Filters on Soliton Interactions in Wavelength-Division-Multiplexing," *Opt. Lett.*, Vol. 17, 1992, pp. 988–990.

[27] Mollenauer, L. F., J. P. Gordon, and S. G. Evangelides, "The Sliding-Frequency Guiding Filter: An Improved Form of Soliton Jitter Control," *Opt. Lett.*, Vol. 17, 1992, pp. 1575–1577.

[28] Suzuki, M., et al., "Reduction of Gordon-Haus Timing Jitter by Periodic Dispersion Compensation in Soliton Transmission," *Elect.. Lett.*, Vol. 31, 1995, pp. 2027–2029.

[29] Doran, N. J., "Dispersion Managed Soliton Systems," *ECOC'98 Proc.*, Vol. 1, Madrid, Sept. 20–24, 1997, pp. 97–99.

[30] Leclerc, O., and E. Desurvire, "Performance Limits in 80–160 Gbit/S (N × 20 Gbit/S) Regenerated WDM Soliton Transmission," *ECOC'97 Proc.*, Vol. 3, Edinburgh, Sept. 22–25 1997, pp. 103–104.

[31] Morita, I., et al., "40 Gbit/s Single-Channel Soliton Transmission over 6300 Km Using Periodic Dispersion Compensation," *OECC'98 Proc., Technol. Digest*, Chiba, Japan, 1998, pp. 326–327.

[32] Shimoura, K., I. Yamashita, and S. Seikai, "Location Effect of Dispersion Compensation Elements in 40 Gbit/S Soliton Transmission Line," *ECOC'2000 Proc.*, Vol. 3, Munich, Sept. 3–5 2000, pp. 259–260.

[33] Georges, T., and F. Favre, "WDM Soliton Transmission in Dispersion-Managed Links: Towards Terabit/S over Megameters," *OECC'98 Proc., Technical Digest*, Chiba, Japan, 1998, pp. 324–325.

3

Dense WDM and Demultiplexers

3.1 Passive Components: The Current Available Choice

3.1.1 Dense WDMs Are Making Optical Network Design Practical

High density wavelength division multiplexing (HDWDM), typically 32 channels at 100-GHz spacing, is widely used in telecommunication optical networks. The commercial availability of very high density wavelength division multiplexing (VHDWDM), typically 96 channels, 50-GHz spacing was announced in 2000 (Figure 3.1). The feasibility of a switchless network at a national scale with 800 wavelengths with a spacing down to a few

Figure 3.1 96 channels, 50-GHz-spacing, athermal dense multiplexer WDM. (*Source:* HighWave Optical Technologies.)

gigahertz is now being studied within a new European research program called ACTS 3rd, SONATA.

Why is there such an explosive growth rate? WDMs multiply network capacity at affordable costs and provide unmatched possibilities for optical routing and optical switching. WDMs also provide signal transparency and large network design flexibility. Some years ago, bit rates of a few Gbps, available on SM fibers without optical multiplexing, seemed to be more adequate. However, they failed to meet the needs of the new multimedia communications appearing now. Today, networks need to deal with the Terabits/s rates required by growing needs for computer networking through private links, TV and HDTV broadcasting, Internet data communications, and video conference on the public networks. Increasing volumes of signals have to be transmitted, switched, processed, and exchanged through the different nodes of multi-Gbps optical systems.

It would be interesting to try to guess how much bandwidth would be necessary to get as much information as is available with the naked eye! Our eyes have more than 200 million rods and cones that encode images with multilevel contrast, in depth and color in a form that can be used by our brain at a rate larger than 25 per second! This means 2*200 000 000*8*3*2 $= 2.310^{11}$ bps, 2,300 Gbps! Considering that images can be transmitted today at only a few hundred Mbps, with advanced multiplexed-broadcasting fiber networks we know that some progress is still necessary.

3.1.2 DWDM Component Technologies

For WDM passive components, there are a number of technologies available with different cost and performance profiles. The main options are:

- Integrated optics arrayed waveguide gratings (AWG);
- Integrated optics concave gratings (CG);
- Fiber Bragg gratings (FBG);
- 3D optics filters (OF);
- 3D optics gratings (OG);
- Cascaded Mach-Zehnders;
- Other devices.

3.2 AWG

3.2.1 Earlier Research

In 1988, a new "focusing and dispersive planar component based on an optical phased array" was proposed by Smit [1]. It was named analogously with the radio frequency phased-array antenna. (Today several different names are used for this component: phasar, phased array, or AWG. The latter is used more often.) The idea was to design an optical phase distribution similar to distributions used in radar and known to produce narrow beams by interference of elementary equispaced sources with constant dephasing from one to another. Several teams contributed to these devices very early on [1–9]. Among them, Takahashi, et al. [4] proposed to increase the optical path difference between "diffracting" elements using the waveguide structure shown in Figure 3.2. Soon Dragone showed how to design N × N arrayed waveguide multiplexers with a planar arrangement of two star couplers [5–7].

3.2.2 Principles of AWG

The phase difference between two adjacent paths in the waveguide array of Figure 3.3 is constant. If n_c is the waveguide index, λ_v is the wavelength in vacuum, and ΔL is the difference of length between two adjacent waveguides, then the phase difference is:

$$\Delta\varphi = 2\pi n_c \Delta L / \lambda_v$$

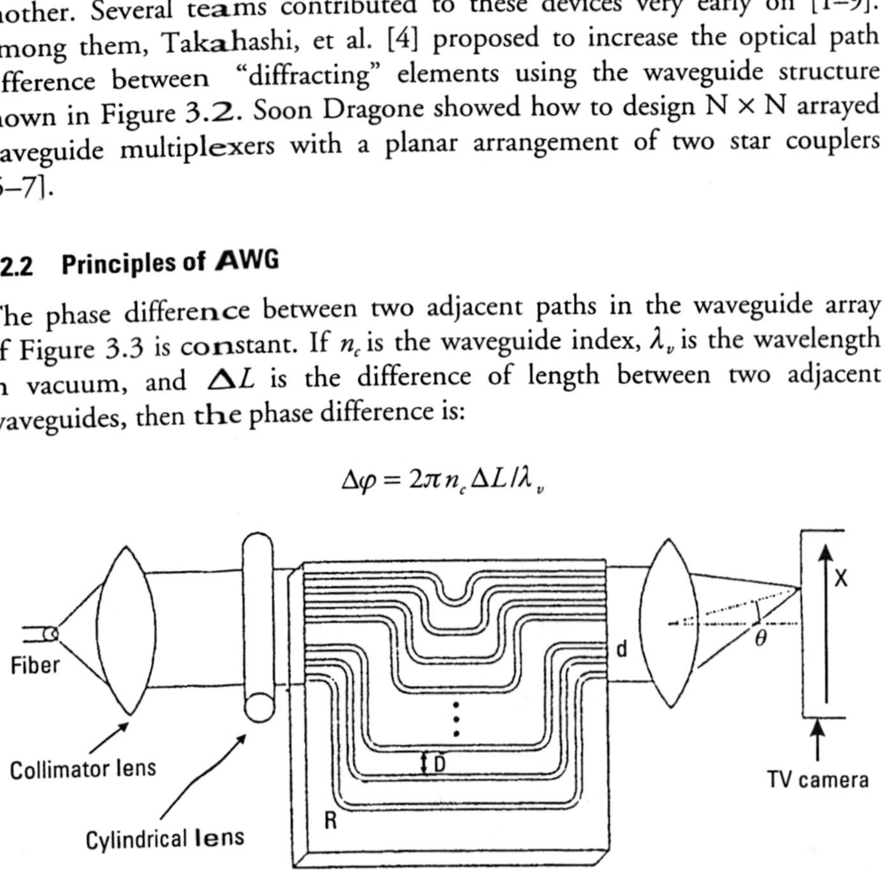

Figure 3.2 AWG with an increased optical path difference between diffracting elements for nanometric resolution. (*Source:* Takahashi, et al. [4].) The length difference between adjacent channels is: $\Delta L = 2(D\text{-}d)$. Transmission occurs for wavelengths such that $n_c\Delta L + n_s d \sin\theta = m\lambda$.

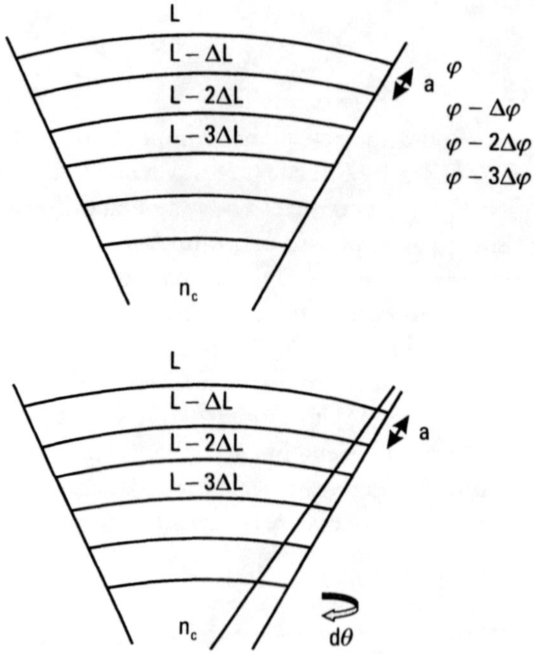

Figure 3.3 AWG principle A.

We get constructive interference and hence maximum transmission when $\Delta\varphi = m\,2\pi$ (where m is an integer, the so-called diffraction order) on the axis of the coupling slab (Figure 3.4) for:

$$\lambda_{v0} = n_c \Delta L / m \tag{3.1}$$

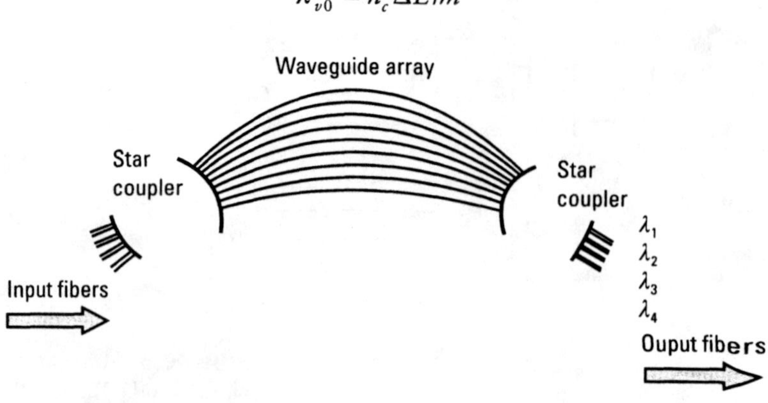

Figure 3.4 AWG principle B.

For all passive devices, silica on silicon or polymer waveguides can be used. Silica or polymers are not suitable for the design of very small devices; due to a low optical contrast, the bending radii cannot be made smaller than several mms without redhibitory losses. For active integrated devices, III–V materials on which radii can be much shorter (typically 500μm) are generally preferred.

3.2.3 Dispersion

For a wavelength variation $d\lambda$ from λ_{v0}, the optical path difference between two adjacent waveguides varies according to:

$$d\Delta L = d(m\lambda / n_c) = m\left(d\lambda / n_c - (\lambda / n_c^2)dn\right)$$

We then get a variation of direction of the wavefront $d\theta$:

$$\begin{aligned} d\theta &= (m/a)\left(d\lambda / n_c - (\lambda / n_c^2)dn\right) \\ d\theta / d\lambda &= (m/a) / n_c - (m/a)(\lambda / n_c^2)(dn/d\lambda) \end{aligned} \quad (3.2)$$

In this equation, we can replace m with $m = n_c \Delta L / \lambda_{v0}$ given by (3.1). And at a distance f from the waveguide grating exit we get a position shift of the maximum: $dx = \mathrm{f}\, d\theta$. Then we get the dispersion at the focus of the second slab (with an index n_s) in front of the fiber exit:

$$\Delta x / \Delta\lambda = (\mathrm{f}\Delta L)\left(n_c - \lambda_{v0}(dn_c / d\lambda)\right) / (\mathrm{a} n_s \lambda_{v0}) \quad (3.3)$$

If the distance between two adjacent fibers is $\Delta x = \mathrm{D}$, the wavelength channel spacing is

$$\Delta\lambda = \mathrm{a} n_s \lambda_{v0} \mathrm{D} / \left(\mathrm{f}\Delta L\left(n_c - \lambda_{v0}(dn_c / d\lambda)\right)\right) \quad (3.4)$$

3.2.4 Free-Spectral Range

We have seen that the wavelength center is $\lambda_{v0} = n_c \Delta L / m$. In AWG, the diffraction order m is generally large (typical mean value $m = 60$ on 32 channels, 100-GHz spacing AWG). For different values of m we get different wavelengths on the same channel. The minimum distance between these wavelengths is obtained for $\Delta m = +- 1$. If we consider in a first approximation that $n_c \Delta_L$ is a constant with respect to the wavelength variation around $\lambda_{\nu 0}$, then:

$$m\Delta\lambda_{\nu} + \lambda_{\nu 0}\Delta m = 0$$

For $\Delta m = \pm 1$ we get the free spectral range:

$$\Delta\lambda_{\nu} \approx \lambda_{\nu 0}/m \qquad (3.5)$$

On the same fiber we get a set of almost equidistant wavelengths separated by $\lambda_{\nu 0}/m$. This interval is called the free spectral range (λ_{FSR}).

Of course in the frequency domain, the corresponding free frequency range is:

$$\nu_{\text{FSR}} \approx \nu_{\nu 0}/m$$

3.2.5 Free Spatial Range and the Number of Available Channels

Let us define the free spatial range X_{FSR} as the distance between two adjacent diffraction orders $m \pm 1$ in the spectrum in front of the exit fibers.

From (3.1), (3.3) and (3.5) we get the free spatial range:

$$X_{\text{FSR}} \approx \lambda_{\nu 0}\text{f}/(n_s \text{a}) \qquad (3.6)$$

With a distance D between exit channels in the spectrum we can use N channels

with:

$$X_{\text{FSR}} = N\text{D}$$

So the number of channels is:

$$N \approx \lambda_{\nu 0}\text{f}/(n_s \text{aD})$$

3.2.6 Polarization Dependency

Some polarization dependency arises from the birefringence of the waveguides. This birefringence, depending on the material used, is high with InP–based substrates, but lower with silica waveguides. In SiO_2 on Si waveguides, it is necessary to take into account the residual birefringence resulting from the compressive stress caused by the different thermal expansion of the silica layer and the silica substrate. The defect can be reduced using an additional top layer of amorphous Si [8] on the SiO_2 upon Si waveguides.

It was also proposed to use a half-wave plate in the center of the waveguide array for a TE to TM mode conversion such that the birefringence effect on the polarization in the first half of each waveguide is in principle exactly compensated in the second half [9]. However, this induces a few decibels of additional losses.

Another solution proposed for birefringence compensation was to match the TE to TM shift with an order shift from m to $m + 1$. However, this works for a limited number of channels only.

The polarization effect can be canceled by special waveguide designs [10] such as rectangular raised strips using "quasi-InP" (InGaAsP λ_g = 1.0 μm) on an InP substrate. An improved technology for eliminating the polarization dispersion in InP phasars using a new compensating waveguide was given in [11]. A 16-channel, 3.2 nm-wavelength-spacing chip with losses as low as 1.3 and 1.7 dB for TE and TM polarizations, respectively, was obtained. The polarization dispersion was smaller than 0.2 nm.

3.2.7 Thermal Drift

We saw that the maximum transmission in the central channel corresponds to a wavelength:

$$\lambda_{\nu 0} = n_c \Delta L / m$$

Both n_c and ΔL vary with temperature.

$$d\lambda_{\nu 0} / \lambda_{\nu 0} = dn_c / n_c + d\Delta L / \Delta L$$

$$d\lambda_{\nu 0} / dT = \lambda_{\nu 0} \left((1 / n_c) dn_c / dT + (1 / \Delta L) d\Delta L / dT \right)$$

$(1/\Delta L) d\Delta L / dT$ is the thermal expansion coefficient ε.

$$d\lambda_{\nu 0} / dT = \lambda_{\nu 0} \left((1/n_c) dn_c / dT + \varepsilon \right) \tag{3.7}$$

With silica waveguides: $d\lambda_{\nu 0}/dT \approx 0.012$ nm/°C at 1,550 nm.

This gives in frequency: $d\nu_{\nu 0}/dT \approx -1.5$ GHz/°C.

This frequency drift is relatively important and must be controlled with a heater or a Peltier device requiring a few watts' power consumption. An athermal AWG operating in the 0 to 85°C temperature range was demonstrated [12]. The temperature dependence of the optical path in silica is compensated with a triangular groove filled with silicone adhesive which has a highly negative thermal coefficient compensating for the small positive temperature coefficient of the silica. The silicone groove works as a microprism inserted between the different silica waveguides, adding optical path proportional to the original path difference along each waveguide. For an optical path *nL* (*n* being the index and *L* the distance) the variation of *nL* with the temperature T is:

$$[d(nL)/L]/\ dT = -\ 37 \times 10^{-5}\ (1/°\text{C}) \quad \text{for silicone adhesive}$$

$$[d(nL)/L]/\ dT = +\ 1 \times 10^{-5}\ (1/°\text{C}) \quad \text{for silica}$$

Consequently an optical length difference ΔL between silica waveguides can be compensated by a length of silicone adhesive that is 37 times smaller.

3.2.8 Typical Values

More theoretical considerations and experimental results were given in [13].

For instance, for 32, 64, and 128 channels, respectively, at 100-, 50-, and 25-GHz spacing, the author gives the following values: same path difference and diffraction order $\Delta L = 63\ \mu$m and $m = 59$, and respectively, f = 11.35, 24.2, and 36.3 mm, number of arrayed waveguides respectively, 100, 160, and 388, crosstalk respectively, <–28, <–27 and <–16 dB.

3.2.9 Technical State of the Art

It is difficult to design AWG with many channels without redhibitory crosstalks. However, cascading AWGs permit increases to the channel number with acceptable crosstalks but with increased losses. In 1999, a component with 320 channels at 10-GHz spacing in the 1,550-nm region was reported [14]. It is an association of a 100-GHz-spaced AWG with 10-GHz-spaced subsidiary AWGs. The crosstalk was < –20 dB. In 2000, industrial AWGs are available with up to 64 channels with 50-GHz spacing in the C band, and up to 40 channels with 100-GHz spacing in the L band. Typical insertion loss values < 5 dB and crosstalk smaller than –30 dB are claimed [15]. One of the main problems to be solved is the dispersion resulting from phase and amplitude errors arising during manufacturing of the gratings [16].

3.3 FBG

3.3.1 Periodic Modulation of Index in the Fiber Core

Fiber photosensitivity was first observed in germanium-doped silica fibers by Hill and coworkers in 1978 [17, 18]. Under long exposure of germanium-doped fiber cores with intense 488-nm argon laser light they observed a back reflection increasing with exposure time. With spectral measurements they understood that this effect was due to a permanent periodic refractive index. This "grating" effect was photoinduced along the core from a weak-standing wave pattern coming from interference between the incident beam and its reflection from the cleaved end of the fiber. Up to 90% reflectivity at the exposure wavelength was obtained, corresponding to an index modulation estimated to be 10^{-5} to 10^{-6}. Later, it was shown that the index variation depended on the square of the writing power [19] and that the effect was rather general and could be observed on many different types of fibers. Much larger index changes were reported. Germanium-boron codoping gave saturated index changes much larger than those obtained in pure germanosilicate (see Table 3.1). Other codopants and other fiber types were used. More practical methods of index writing were proposed. Recording from the side of the fiber in the interference pattern of two coherent beams, or through a phase mask allowed the recording of any period, and made reflectance at any wavelength with predictable spectral shape possible. The sources are generally UV lasers (e.g., 248- and 193-nm excimer lasers), or 244-nm argon lasers with intracavity doubling. However, 157-nm F_2 excimer lasers give highly efficient photosensitivity in germanosilicate fibers. Also, less efficient 334- or 351-nm lasers have been used. New excimer krypton fluoride lasers with an unstable resonator, exhibiting high UV output at 248 nm, ideal for Bragg grating writing, became available in 2000. The use of an unstable resonator, on large-volume gain media such as excimer, gives a larger spatial coherence with a reduced divergence. However, this reduces the laser intensity to amplified spontaneous emission ratio. A cylindrical unstable resonator gives expansion of the beam along an axis parallel to the fiber with laser gain losses reduced by several orders of magnitude when compared with a spherical unstable resonator giving an equivalent magnification. A spatial coherence larger than 1mm was obtained [20]. Today, as well, CW frequency-doubled ion lasers giving 500-mW CW at 244 nm are available (as claimed by the Coherent Laser Group). They allow fast integration into either a phase mask or holographic fiber-writing apparatus.

Hydrogenation of the fiber prior to writing and flame brushing also enhance photosensitivity [21]. However, the long-term stability of components made with excessive flame brushing could be an issue. Sn-Ge: SiO2 germanosilicate fibers give better photosensitivity and thermal stability than Ge: SiO_2 fibers (with or without boron doping). Many of the developments have concentrated on germanosilicate fibers (Table 3.1), but of course, photosensitivity is not restricted to them.

3.3.2 Main Properties of FBG

The light guided along the core of an SM fiber normally travels as a plane "inhomogeneous" wave perpendicular to the axis: This corresponds to the "single mode" of propagation. Most of the power travels in the core with a small part corresponding to the evanescent wave that travels outside near the

Table 3.1
Photosensitivity with or Without Boron Codoping in Germanosilicate Fibers (Writing 1 W/cm^2 Doubled CW Ar Laser)

Fiber Type	Fiber Δn	Saturated Index Modulation	Maximum Reflectivity for 2 mm Gratings (%)	Time for Reflectivity to Saturate
Standard ~4 mol% Ge	0.005	3.4×10^{-5}	1.2	2 hours
High index ~20 mol% Ge	0.03	2.5×10^{-4}	45	~2 hours
Reduced fiber ~10 mol% Ge	0.01	5×10^{-4}	78	~1 hour
Boron codoped ~15 mol% Ge	0.003	7×10^{-4}	95	~10 min

(*After:* Williams, et al. [22].)
Note: An exhaustive analysis of FBG is given in [23].

core in the cladding. This wave is perturbed by the constant index-grating planes perpendicular to the axis (Figure 3.5). A weak reflection at each grating plane takes place. The contributions of each reflection add constructively in the backward direction for wavelengths defined by the grating period d, the order *m* (an integer 1, 2...). And the effective index in the core n_{eff}:

$$\lambda_B = 2n_{eff}\,\mathrm{d}/m \tag{3.8}$$

From coupled mode theory [19] the reflection $R(\lambda, L)$ at the maximum is:

$R(\lambda, L) = \tanh^2(\Omega L)$ where L is the grating length and $\Omega = \pi \Delta n\, M_p/\lambda$.

M_p is the fraction of mode power in the core. For constant parameters along the grating, and without saturated reflection, the spectrum is a sinc function of λ: We get a Fourier transform relationship between L and λ. When L becomes larger the spectral width of the reflection becomes smaller. The FWHM (function width at half maximum) [from 24] is:

$$\Delta\lambda \cong \lambda_B s \sqrt{\left[(\Delta n/2n_0)^2 + (1/N)^2\right]} \tag{3.9}$$

Where N is the number of periods and s is a parameter varying between 0.5 for weak reflections to 1 for strong reflections.

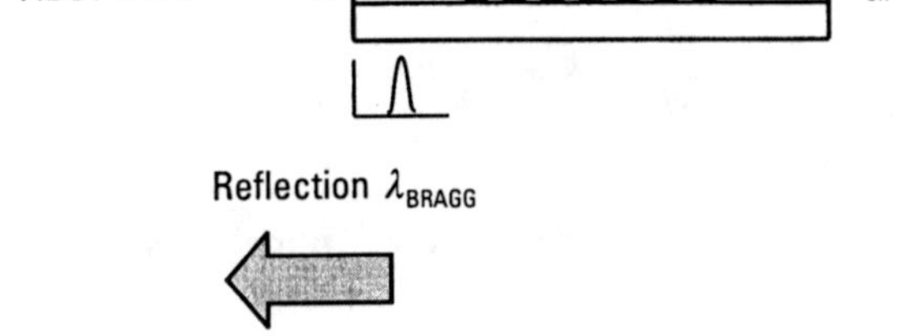

Figure 3.5 FBG.

However, in case of saturated reflection, the reflection spectrum broadens and becomes "flat top." The grating period d(z) can be varied monotonically along the core, so that the Bragg wavelength λ_B varies along the grating:

$$\lambda_B(z) = 2n_{eff}\mathrm{d}(z)/m$$

This method is used for manufacturing WDM components with broadband flat-top channels. Of course, such components are more tolerant to laser wavelength drifts.

Apodization of the reflection spectrum can be obtained by varying the coupling along the grating, weighting the modulation amplitude. This can be done using a phase mask [25, 26] or other solutions, which modify the diffraction efficiency along the grating. This method is especially effective for side mode suppression. With a particular refractive index correction (modulation of the refractive index varying as a Gaussian profile along the grating around a mean value), symmetric transmission profiles with side mode suppressed are obtained [27].

Blue-side loss: Normally, a highly reflective BG filter made from a standard SM fiber shows undesirable losses in a wavelength region a few namometers under its Bragg wavelength. It was shown that this defect is due to the coupling to cladding and leaky modes by diffraction of the LP01 mode at the boundary between core and cladding. The LP01 mode field, that is larger than the core, is perturbed at the interface of the grating and is limited in the core. Different methods exist to reduce this effect. A very efficient method is to expand the photosensitive area in an additional Ge + F-doped inner cladding. The first experimental results of blue-side loss attenuation by this method were reported by E. Delevaque, et al. [28]. Moreover, the mode field diameter can be controlled as the refractive index changes with the fluorine concentration. So the splicing losses between the FBG and a standard SM fiber (G.652 or other) can be minimized to almost negligible values.

3.3.3 Different Types of Bragg Gratings

Bragg gratings are generally classified into several types:

- Type I: Grating written in classical photosensitive fibers with moderate powers. They can be used in the –40°C to +80°C temperature range necessary in telecommunications but are erased at about 200°C. They have small losses.

- Type IIA: Grating written with relatively long exposure duration. The main cause of index variation corresponds here to glass compaction. They can be used at higher temperatures. They are erased at about 500°C.
- Type II: Grating written with single high-power pulse (>0.5 J/cm^2). They couple light into the cladding at $\lambda < \lambda_B$. They can be used at higher temperatures (800°C to 1,000°C), but their mechanical reliability can sometimes be an issue.

For WDM and DWDM components type I is generally preferred. Several gratings with different periods can be written at different places along the fiber or superimposed at the same location, so that multiwavelength components can be made.

3.3.4 Drift with Temperature

From derivation of (3.8) we get:

$$d\lambda_B/dT = \lambda_B\left(\left(1/n_{eff}\right)dn_{eff}/dT + \varepsilon\right) \tag{3.10}$$

In which ε is the thermal expansion coefficient.

With germanium-doped silica fibers:

$$d\lambda_B/dT \approx 0.014 \text{ nm/°C at } 1{,}550 \text{ nm.}$$

This gives in frequency:

$$d\nu_B/dT \approx -1.75 \text{ GHz/°C.}$$

The main drift comes from the variation of index dn_{eff}/dT. This drift can be compensated by mechanical length compensation in a package made of two materials with different thermal expansions.

3.3.5 Typical Specifications of Available Bragg Grating DWDM

Some typical specifications of commercially available products are given in Table 3.2 and in Figures 3.6 and 3.7. These DWDM components are thermally compensated. The residual wavelength shift with temperature is smaller than 50 pm/80°C.

Table 3.2
Typical Specifications of FBG

Specifications ITU (GHz)	50	100
–0.5 dB-bandwidth (nm)	0.15	0.3
–15 dB-bandwidth (nm)	0.4	0.5
Inband isolation (dB)	30	30
Outband isolation (dB)	25	25
Package size (mm)	62 × 7 × 9.5	32 × 7 × 6.5

(*After:* E. Delevaque and D. Pureur [29]).

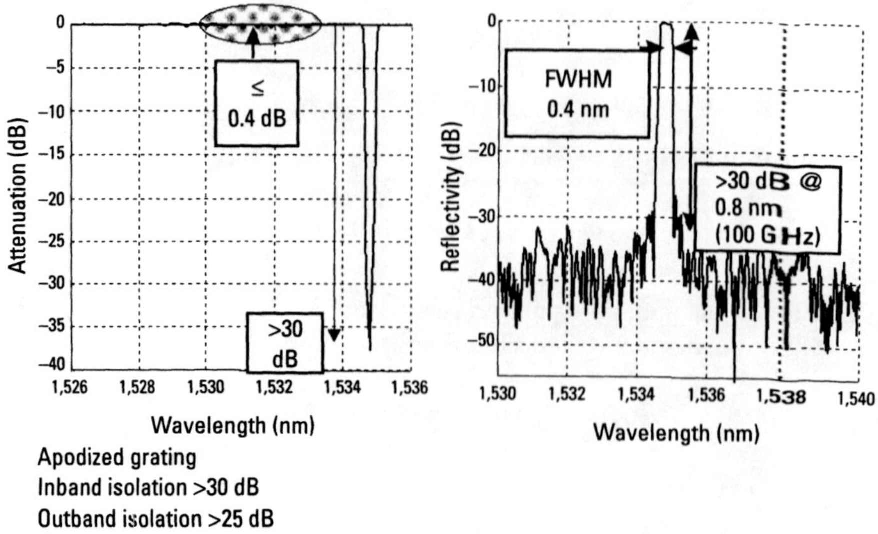

Figure 3.6 Typical transmission FBG 100-GHz spacing. (*Source:* HighWave Optical Technologies.)

3.4 Optical Multidielectric Filters

3.4.1 General Principles

3.4.1.1 Long-Wavelength and Short-Wavelength Pass Filters

Multidielectric filters produce a light-beam angular separation in reflecting a given spectral range and transmitting the complementary part. These two spectral ranges can be very large, particularly with current edge filters: long-

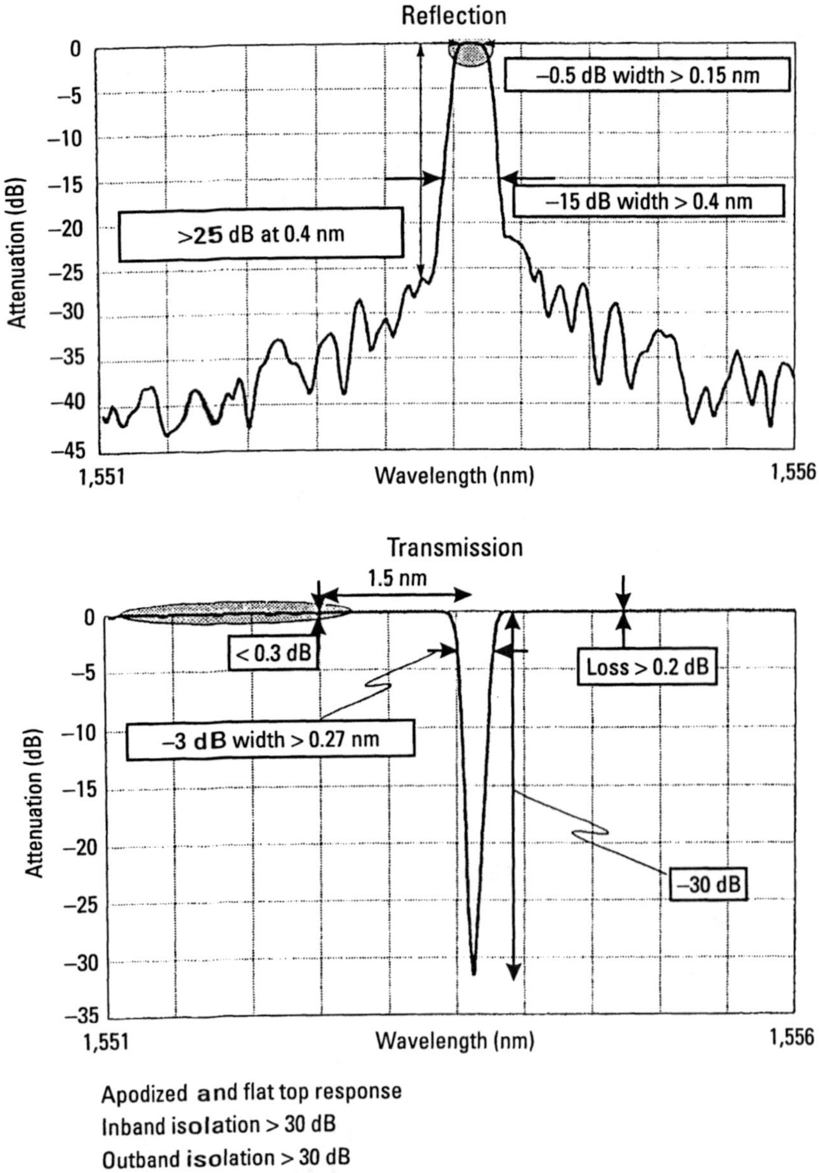

Figure 3.7 Typical transmission FBG 50-GHz spacing. (*Source:* HighWave Optical Technologies.)

wavelength pass filter (LWPF) and short-wavelength pass filter (SWPF). These filters consist of stacks of alternatively high (H) and low (L) index layers on a substrate (S). Each layer has an optical thickness such that $ne = \lambda_0/4$ in order 0 filters and $ne = 3\lambda_0/4$ in order 1 filter. Often, stack structures $(H/2\ L\ H/2)^K$ [30] are used and the main problem is to obtain sharp-edge and high-reflectivity power filters (R > 99%) in a given spectral range, and simultaneously to obtain a good transmission in the complementary spectral range (T > 99%). If the (H/2 L H/2) elementary sequence is rigorously repetitive, we obtain reflection curves such as those shown in Figures 3.8 and 3.9, with oscillations on edges that can be corrected by an admittance adaptation on the first and last few layers; unfortunately, this is detrimental to the relative sharpness of the transmission edge.

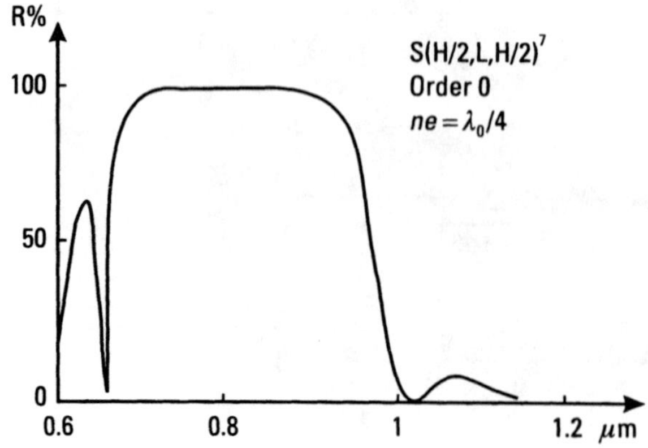

Figure 3.8 LWPF theoretical reflection curve without adaptation.

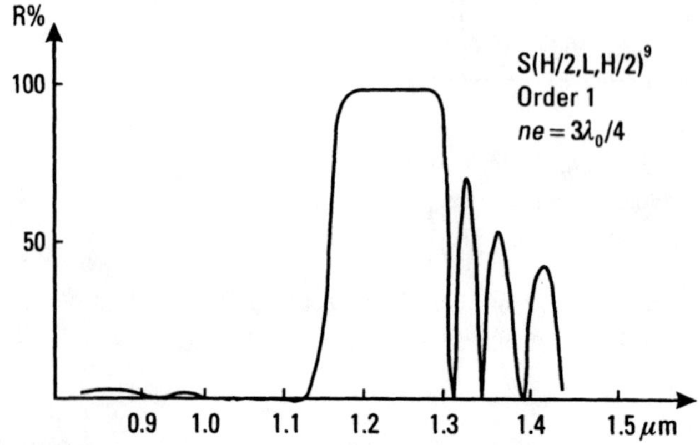

Figure 3.9 SWPF theoretical reflection curve without adaptation.

In Figure 3.10, LWPF theoretical and experimental transmission curves (31 layers), in the "first window" for 0.82/0.85 μm separation, of a filter consisting of a stack of modified thickness layers are presented. In Figure 3.11, the same curves and the theoretical curve of an equivalent filter with nonmodified layers are shown for 1.3/1.5 μm separation.

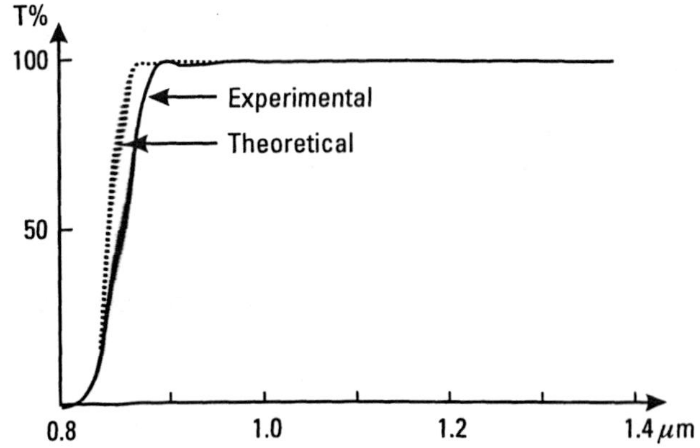

Figure 3.10 LWPF theoretical, and experimental, curves $(H/2\ L\ H/2)^{15}$ with adaptation.

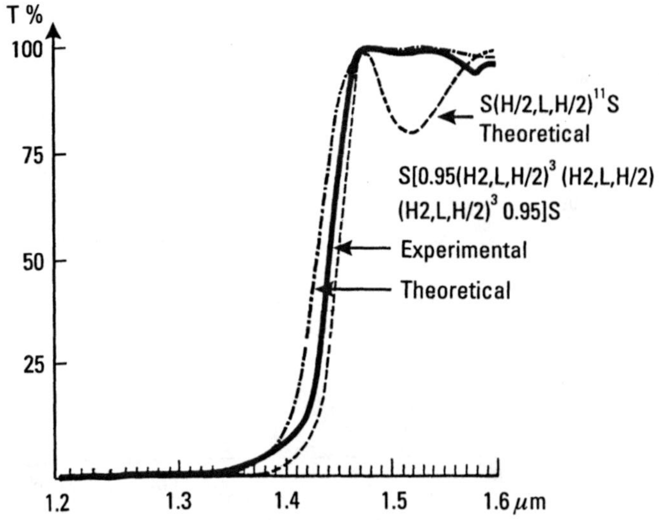

Figure 3.11 1.3/1.5 μm filter transmission curves, theoretical without adaptation, theoretical with adaptation, experimental with adaptation.

The manufacturing of SWPF is more delicate, as order 1 filters ($ne = 3\,\lambda_0/4$) with corrected entrance and exit layers are again required, and, consequently, the total coating is thicker. In any case, stringent control rules must be applied with longer deposition durations in the SWPF case. In practice, a spectral separation, $\Delta\lambda = 0{,}02\,\lambda$ between channels can be obtained, but to take into account possible manufacturing defects, we prefer to recommend a larger spectral separation, $\Delta\lambda > 0{,}05\,\lambda$. It must also be pointed out that the thermal stability of traditional filters is 2×10^{-2} to 10^{-1} nm/°C (for enhanced performance see below).

3.4.1.2 Narrow Bandpass Filters

For multichannel dense WDM, a third filter type is often used in devices in which the multiplexing (or demultiplexing) is carried out through successive injections (or extractions) by transmission of a thin spectral band and reflection of the complementary spectral domain. It consists of multiple cavity filters used as bandpass filters (BPF). In Figure 3.12, theoretical and experimental reflection curves obtained with a stack of 23 layers and 3 cavities with TiO_2 and SiO_2, respectively, high and low index layers are shown [30–33]. Using more cavities allows a dramatic bandpass narrowing that is necessary for dense multiplexing. The bandpass width is reproducible within few percentages (10% to 20% difference is typical between different batch-processings; however, a better specification can be obtained at higher cost).

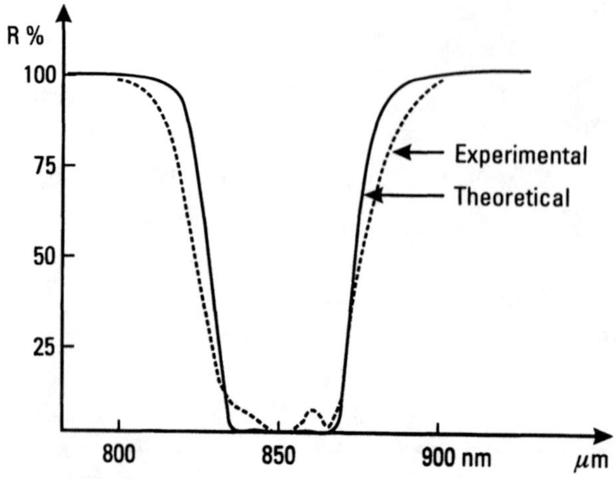

Figure 3.12 Theoretical and experimental curves of a 23-layer multicavity BP filter (high index 2.45, low index 1.47, substrate index 1.563).

The center wavelength is also quite sensitive to process parameters but can be adjusted later, if necessary, with a change of the incidence angle. The wavelength shifts almost linearly at low incidence angles.

3.4.2 Materials and Processes for Enhanced Performances

Conventional filters are manufactured with Zinc sulfide (ZnS) or Zinc selenide (ZnSe) and cryolite. However, these coatings are hygroscopic and they must be protected from the environment with an epoxy housing. They are relatively cheap, as the manufacturing is easy. The evaporation can be made from simple filament-heated crucible sources. But they suffer from poor temperature stability and low transmission. They need to be used with Peltier thermal housing or other temperature controllers adding cost, size, and power requirements. For ZnS/cryolithe made with a conventional thermal evaporation the thermal drift is about 20 pm/°C.

Of course this is not fully compatible with adequate DWDM devices. Metal oxides ("hard coatings") have much improved temperature stability. They are generally evaporated in vacuum chambers through an electron-beam gun heating. But the evaporated films generally grow in columns corresponding to porous and stressy films. These coatings are very sensitive to humidity-inducing wavelength shifts. New methods such as ion beam sputtering, reactive ion plating, and MicroPlasma (Corning OCA patented method [34]) produce a tightly packed structure preventing water from entering the metal oxide layers.

The theoretical thermal shift is:

$$\Delta\lambda = \lambda\left[\varepsilon + (1/ndn)/dT\right]\Delta T$$

Typical temperature dependence of ion-assisted deposited coating at 1.5 μm is less than 0.003 nm/°C. This corresponds to a lower than 0.3-nm drift in a −20°C to +80°C operating temperature range [35].

Very sharp cut-off characteristics can be obtained adding a feedback route to the resonators [36]. It is desirable to get flat passband and wide stop-band of such multi–FP filters. This can be done by adjustment of the different cavity lengths: Using this technique, a 4-THz stop-band and 28-GHz passband was claimed in the 1,550-nm wavelength region by H. Furuta, et al. [37]. Stop-band could be further expanded by using optical directional coupler type multiresonators [38].

3.4.3 Practical Narrow Bandpass Filters DWDM

Among others, typical results for DWDM using filters with up to 150 quarter-wavelength layers and up to 9 optical cavities on practical micro-optics devices were reported in [39]. The typical performance of a component with 16 channels, 100-GHz spacing was: bandwidth at −0.5 dB ≥ 0.22 nm, insertion loss < 6 dB, isolation between adjacent channels ≥ 22 dB, PDL < 0.3 dB, temperature range 0°C to 50°C. On commercially available 8 channels, 200-GHz products, typical specifications could be for example: bandwidth at −0.5 dB ≥ 0.5 nm, insertion loss < 4 dB, isolation between adjacent channels: ≥ 25 dB, PDL: 0.1 dB, temperature range: 0°C to −65°C. Narrow filters for a few 50-GHz-spacing channels can be made.

3.5 Diffraction Gratings

3.5.1 Introduction

Wavelength division multiplexers using filters cannot be used when the number of channels is too high or when the wavelengths are too close. The main advantage of the grating is the simultaneous diffraction of all wavelengths, so that it is possible to construct simple devices with a large number of channels.

A diffraction grating [40–49] is an optical surface, which transmits or reflects light and on which a large number of grooves N_0 (several tens to several thousands per millimeter) are ruled by a diamond tool or obtained by holographic photoetching. The grating has the property of diffracting light in a direction related to its wavelength (Figure 3.13). Hence an incident beam with several wavelengths is angularly separated in different directions. Conversely, several wavelengths λ_1, λ_2, ... λ_n coming from different directions can be combined in the same direction. The diffraction angle depends on the groove spacing and on the incidence angle.

In Figure 3.14, let us consider a transparent and equidistant slit array and an incident plane wave at an angle with the perpendicular to the grating. Each slit diffracts light in transmission. The optical path difference dispersion and resolution are given in Table 3.3.

In the direction α', measured from the perpendicular to the grating, the waves coming from the different slits are in phase if the path difference Δ_0 between the successive optical paths (L_1 M_1) (L_2 M_2) is:

$$\Delta_0 = d(\sin\alpha + \sin\alpha') = k\lambda \tag{3.11}$$

Table 3.3
Order Position–Dispersion–Grating Resolution

Periodicity: d
Optical path difference:
$\Delta_0 = d(\sin\alpha + \sin\alpha') = k\lambda$
Dispersion: $\frac{d\alpha'}{d\lambda} = \frac{k}{d\cos\alpha'}$
Resolution: $R_S = kN_0$

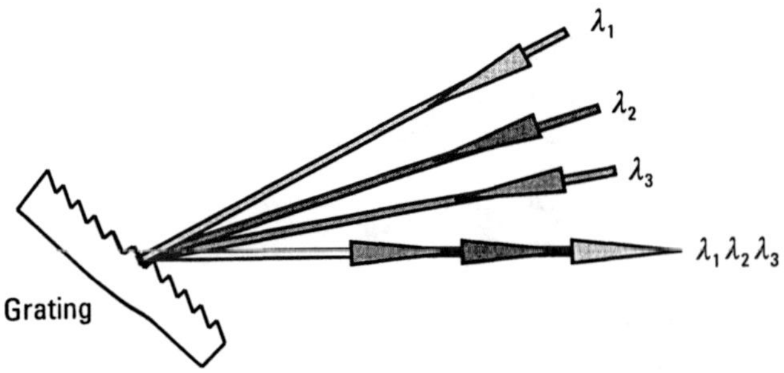

Figure 3.13 Principle of multiplexing by diffraction on an optical grating: Wavelengths $\lambda_1, \lambda_2, \lambda_3$ coming from different directions are diffracted in the same direction into a single transmission fiber.

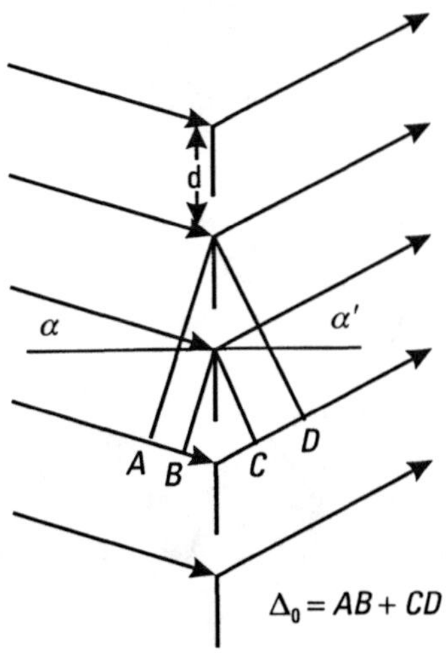

Figure 3.14 Calculation of the diffraction order angles.

k is an integer, λ is the wavelength, and d the distance between two successive slits. $k = 0$ corresponds to direct transmission, $k = \pm 1$ corresponds to the first diffraction orders on each side of the direct transmission.

It is easily demonstrated that the angular dispersion, corresponding to the wavelength variation is:

$$\frac{d\alpha'}{d\lambda} = \frac{k}{\mathrm{d}\cos\alpha'}$$

Consequently, the wavelengths can be angularly separated. The limit corresponds to the angular width of the diffraction of the whole surface of the grating projected in the α' direction. One can show that the maximum resolution that can be obtained is:

$$R_s \max = \frac{\lambda}{d\lambda} = kN_0$$

Where N_0 is the total number of grooves. N_0 is usually very large; hence very small distances between channels can be obtained. In practice, spacing of 0.5 nm between channels has been obtained with grating multiplexers but the theoretical limit is far from being reached. In spectroscopy, resolution $\lambda / d\lambda = 0.5 \times 10^6$ are now standard with commonly used reflection gratings.

The reflection grating case is described in Figure 3.15. If the index outside the grating is n, the former law becomes:

$$\mathrm{nd}(\sin\alpha + \sin\alpha') = k\lambda$$

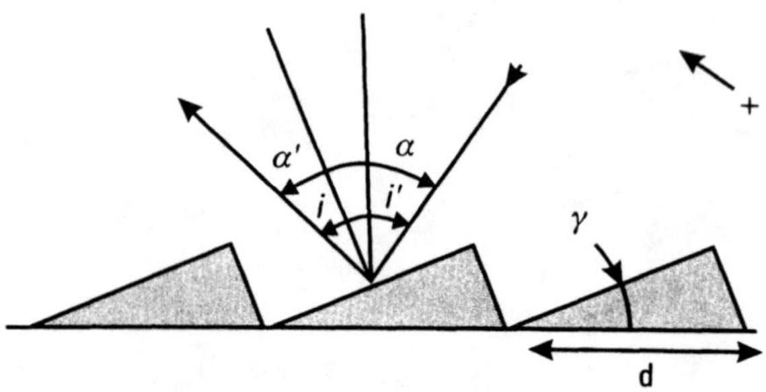

Figure 3.15 Diffraction plane grating.

In which d is the periodic distance between the grooves and α/α' the incident/diffracted angles measured from the perpendicular N to the mean grating surface.

3.5.2 Efficiency Versus Wavelength

The groove shape allows the concentration of the diffracted energy in a given spectral range: The grating is then said to be "blazed."

3.5.2.1 Plane Reflection Grating Study

N: Perpendicular to the mean grating surface;

M: Perpendicular to the facet;

α: Incident angle (from N);

α': Diffracted angle (from N);

i: Incident angle (from M);

i': Diffraction angle (from M);

d: Groove spacing;

γ: Blaze angle.

A. Case in Which the Groove Spacing is Much Larger Than the Wavelength and with Small Angles, Scalar Approximation

We will use a small angle approximation for α, α' and γ and assume that the number of grooves is large enough to have an angular width of the diffraction by the total grating surface much smaller than the angular width of the diffraction by a facet.

We will get rough results, but this analysis is a very useful starting point.

Let us mention that the diffracted energy is maximum in the direction corresponding to a reflexion on each grating facet (i.e., when $i = i'$).

From the relations $i = \alpha - \gamma$ and $i' = \alpha' - \gamma$, we get the blaze angle value

$$\gamma = \frac{\alpha + \alpha'}{2}$$

This γ angle determines the shape of the diamond to be used for the ruling of the grating master.

The grating relation, $d(\sin\alpha + \sin\alpha') = k\lambda$ becomes:

$$2d\sin\gamma\cos\frac{(\alpha-\alpha')}{2} = k\lambda$$

For reflection gratings, the blaze angle is generally calculated in the Littrow conditions, in which $\alpha = \alpha'$, corresponding to an incident and exit beam in the same direction.

In the first order and Littrow condition:

$$\lambda_{1\ \text{blazed}} = 2d\sin\gamma$$

With $\alpha \neq \alpha'$:

$$2d\sin\gamma\cos\frac{(\alpha-\alpha')}{2} = \lambda_{1\ \text{Blazed}}$$

In the second order and Littrow condition:

$$\lambda_2 = d\sin\gamma$$

In the order ξ and Littrow condition:

$$\lambda_\xi = \frac{2d}{\xi}\sin\gamma$$

Each facet of the plane grating gives a diffraction phenomenon that is characterized by the distributed amplitude A:

$$A = A_{\lambda\xi}\,\frac{\sin\dfrac{\pi d}{\lambda}(\sin i + \sin i')}{\dfrac{\pi d}{\lambda}(\sin i + \sin i')}$$

For $\lambda = \lambda_\xi$, the blaze wavelength, the intensity is maximum and equal to:

$$I_{\lambda\xi} = \left(A_{\lambda\xi}\right)^2$$

Let us calculate the ratio:

$$\frac{I_\lambda}{I_{\lambda\xi}} = \Phi\left(\frac{\lambda_\xi}{\lambda}\right)$$

Which represents the spectral distribution of the diffracted intensities. The intensity I_λ corresponding to the wavelength λ is:

$$I_\lambda = I_{\lambda\xi}\left[\frac{\sin\left(k\pi - \frac{\pi\xi\lambda_\xi}{\lambda}\right)}{k\pi - \frac{\pi\xi\lambda_\xi}{\lambda}}\right]$$

For each wavelength $\alpha = \alpha'$, k is the diffracted order considered at λ, ξ is the diffracted order for which the grating is blazed at the wavelength λ_ξ. Figure 3.16 shows the spectral distribution of the diffracted energy in the first order from a grating blazed in the first order ($\lambda_\xi = \lambda_1$).

B. General Case

With gratings having a small groove spacing (d = a few λ or less), the results of section A are no longer valid. Anomalies appear in the efficiency curves and the spectral distribution of the diffracted energy depends on the

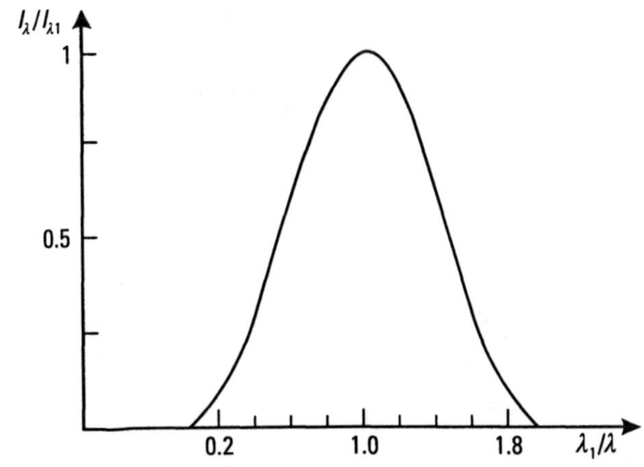

Figure 3.16 First-order Echelette grating efficiency (approximation of small blaze angle). (Normalized abscissa λ_1/λ.)

polarization. The formula $\lambda_{1\ \text{blaze}} = 2\ d\sin\gamma$ is no longer valid. The maximum efficiency with unpolarized incident light is lower than λ_1 given by this formula. The corresponding curves were fully calculated from Maxwell's equations (M. Petit, thesis, Faculté d'Orsay, France, 1966 and [41]). These curves give the efficiencies of perfectly conducting sinusoidal gratings (Figure 3.17) or perfectly conducting triangular profile gratings (Figure 3.18).

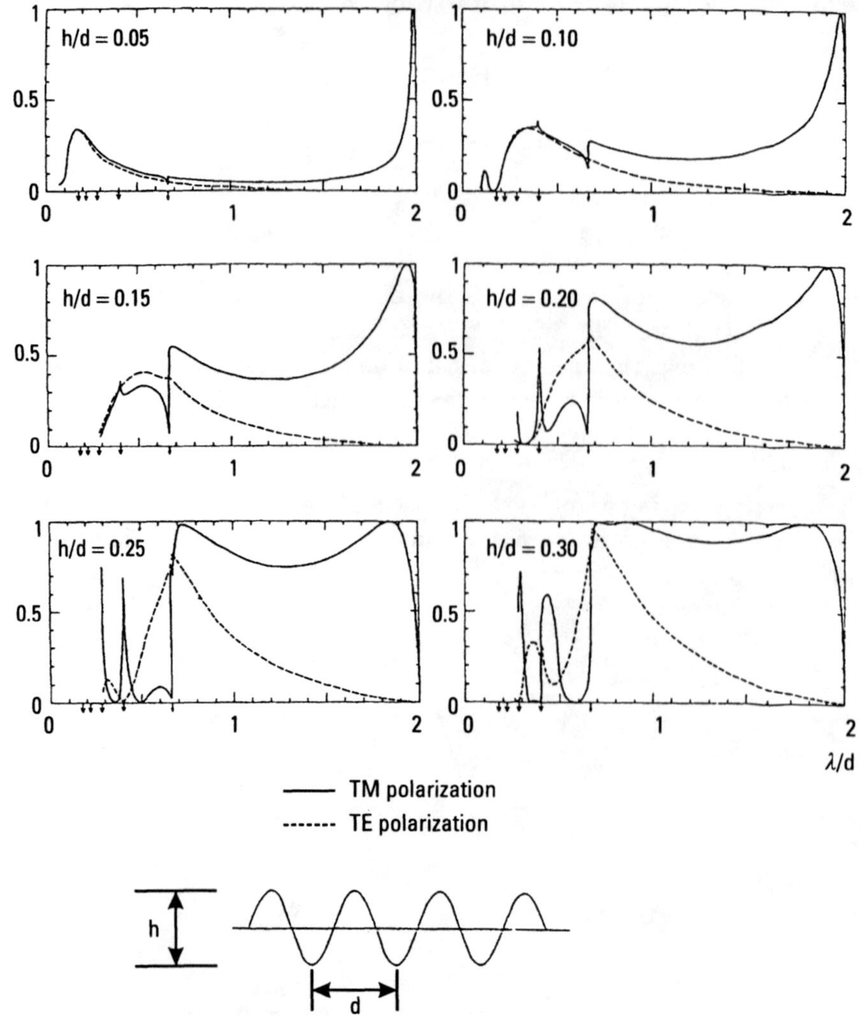

Figure 3.17 Efficiencies as function of the ratio λ/d of sinusoidal profile reflexion gratings with different depths. Electromagnetic theory [41] of perfect conductors.

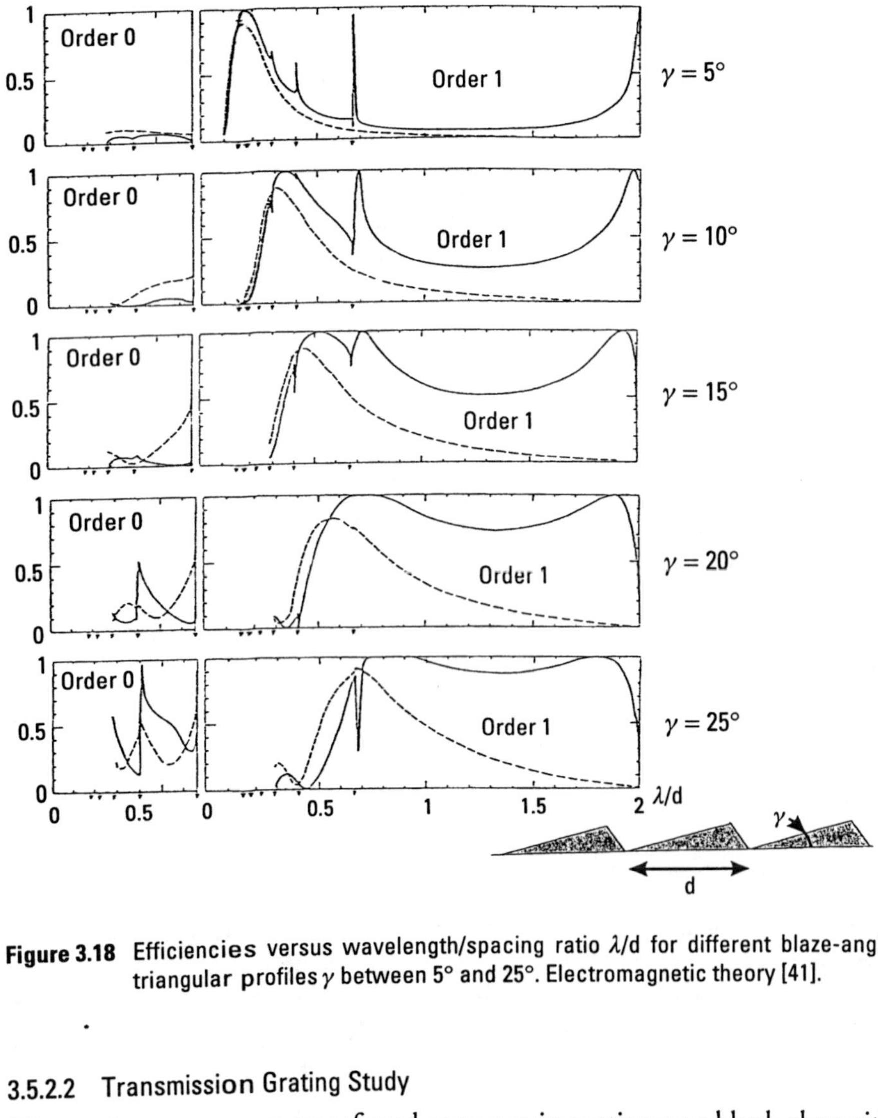

Figure 3.18 Efficiencies versus wavelength/spacing ratio λ/d for different blaze-angle triangular profiles γ between 5° and 25°. Electromagnetic theory [41].

3.5.2.2 Transmission Grating Study

The grating grooves are transferred onto a resin coating on a blank glass with both faces polished within a quarter of a fringe. The grooves can be considered as a set of small diffracting prisms.

Let us consider:

n resin index;

γ facet angle;

d grating period (see Figure 3.19).

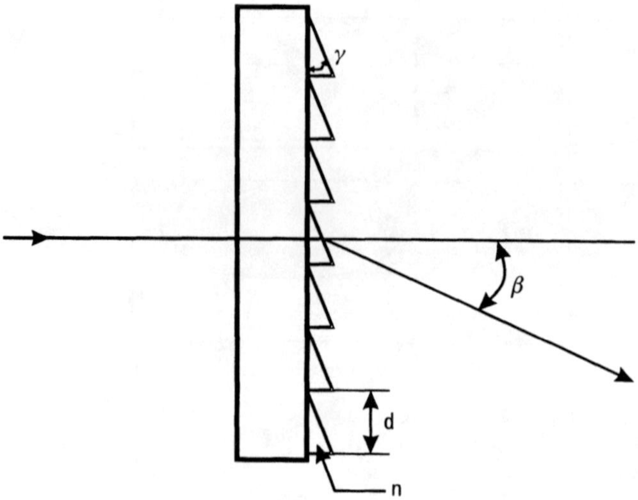

Figure 3.19 Transmission grating.

A. Gratings with Groove Spacing Much Larger Than One Wavelength and Small Blaze Angle (Scalar Approximation)

When the incident light is perpendicular to the blank (Figure 3.19), the blaze wavelength λ_B is given by the formula:

$$\lambda_B = d(n-1)\sin\gamma \tag{3.12}$$

B. General Case

Here again, it is necessary to use the electromagnetic theory to obtain the correct efficiency value. However, it can be demonstrated that one can obtain 100% efficiency by using the formula given above, with a metallic coating on the small facet of the grooves. This is obtained with an electromagnetic field perpendicular to the conductive facet [45].

3.5.2.3 Concave Grating Study

CGs are generally used in reflection; therefore, their blaze angles are calculated like those of plane reflexion gratings. Within the scalar theory, the blaze angle of such gratings has to be changed continuously in order to keep them perpendicular to each facet, bisector of ABC, the angle between incident and diffracted order 1, at all locations of B on the grating surface (Figure 3.20). However, this is not usually necessary, and, in most cases, it is the angle between each facet and the plane tangent at the center P that is kept

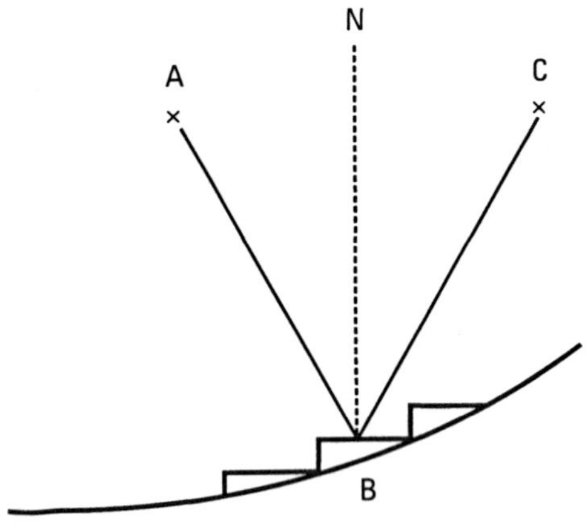

Figure 3.20 Concave grating.

constant. However, on ruled or holographic gratings, a profile variation with surface location can be obtained by variable incidence ion etching. On classically ruled gratings, one can rule three or four zones with a constant but optimized angle, and with a corresponding loss of resolution, in cases in which the efficiency variation would be too large from the center to the edge. The resolution of such a grating ruled in three parts will be at least three times less than that of the same grating ruled as a single part, but, in most cases, it will have an efficiency approximately equal to the efficiency of the equivalent plane grating [46].

The possibility of controlling the focal properties of CGs by a proper distribution of the grooves has been known for a long time. This can be done on computerized ruling engines. However, this is more easily obtained with holographic techniques. During the last 25 years, holographic gratings have evolved drastically. They are capable of stigmatic imaging without the need for auxiliary focusing optics. Ion etching provides a means of blazing and optimizing these gratings. Following production of the holographic master, which has pseudosinusoidal groove profiles, the grating is then used as a mask subjected to an argon laser beam to remove surface atoms until the groove structure presented by the surface hologram is brought into the substrate itself. To "shape" the grooves, the angle of incidence of the ions to the substrate can be adjusted to produce triangular blazed-grating profiles. High efficiencies have been achieved with concave aberration-corrected holographic gratings [47]. Typically 60% to 80% efficiencies and subnanometer wavelength resolution from 1,500 to 1,560 nm on all single-mode concave holographic grating tunable demultiplexers are obtainable.

3.5.2.4 Practical Efficiency

With echelette gratings such as that of Figures 3.21 and 3.22, practically 85% to 90% efficiency with unpolarized incident light is obtained at the blaze wavelength. Polarization by the grating remains small if the blaze angle is small enough (here ~5% on several tens of nanometers). This efficiency is increased by multidielectric coatings [48].

3.5.3 Bandwidth of Grating Devices

3.5.3.1 Devices with Input and Output SM Fibers

For the theoretical analysis, we use the general case of Figure 3.23; we calculate the transmission function from the entrance fiber F_1 towards the exit fiber F_2.

Let us consider a fiber F_1 with a polished end-face in the plane xy perpendicular to the direction zz' axis of the fiber. The focal plane of an optical system L_1 is on the xy plane. The end-face of F_1 is imaged in the focal plane of L_2, $x'y'$, which contains the polished end-face of the fiber F_2, after an angular dispersion by disperser R (generally a grating). xyz and $x'y'z'$ are local right-angled trihedrals.

- We will calculate the energy transmitted through F_2 as a function of the wavelength λ of the light delivered by L_2.

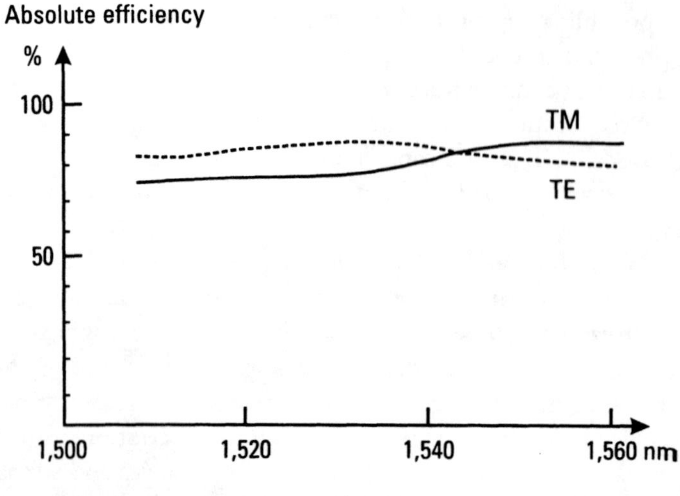

Figure 3.21 Transverse electric (TE) and transverse magnetic (TM) efficiencies measured on a small blaze-angle grating.

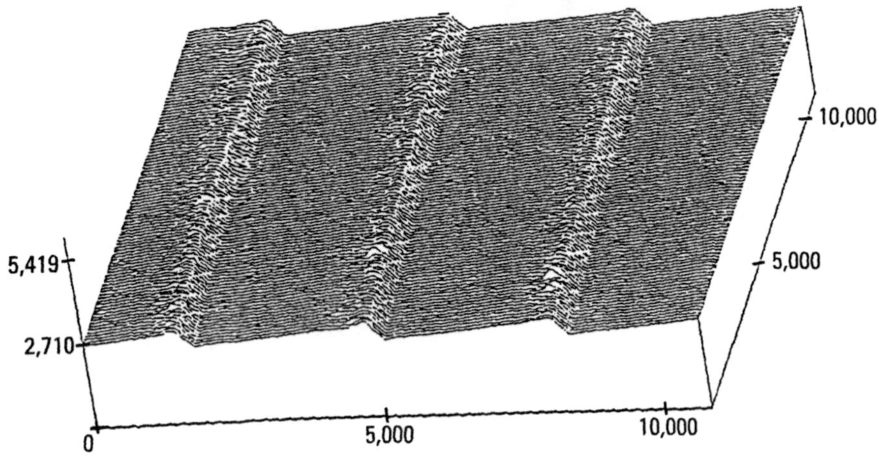

Figure 3.22 Grating profile, tunnel microscope view.

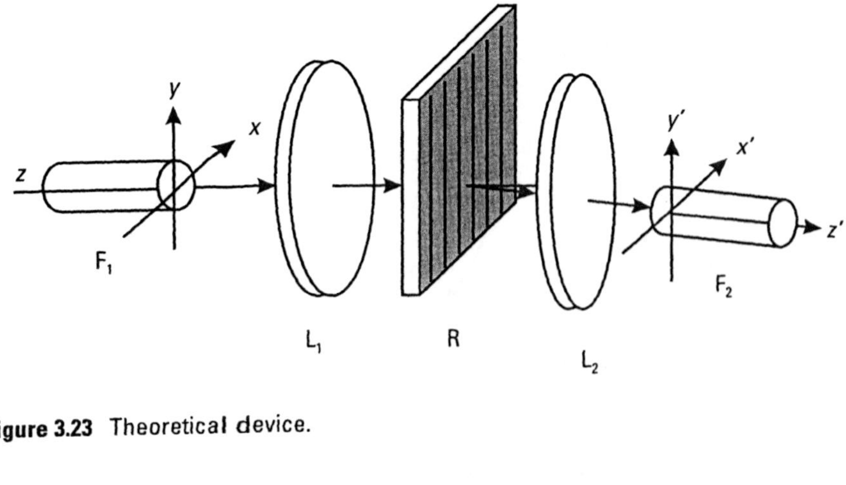

Figure 3.23 Theoretical device.

- We assume hereafter that the magnification of the optical system is 1 and that the couple L_1 and L_2 is afocal. As a matter of fact, that condition is necessary to avoid a coupling loss. (An afocal coupling with a unit magnification keeps an identity between entrance and exit angles. This is the case with the Stimax configuration [see Figure 3.36], but this is not always verified with other configurations.)

- We show what happens in the plane $x'y'$ on Figures 3.24 and 3.25.

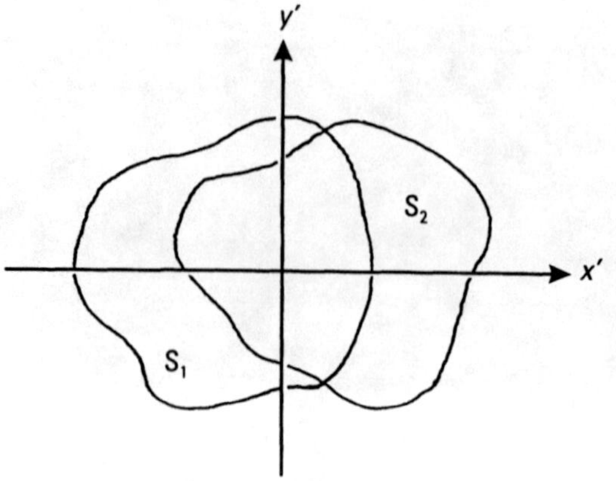

Figure 3.24 Transmission in the exit plane, general case.

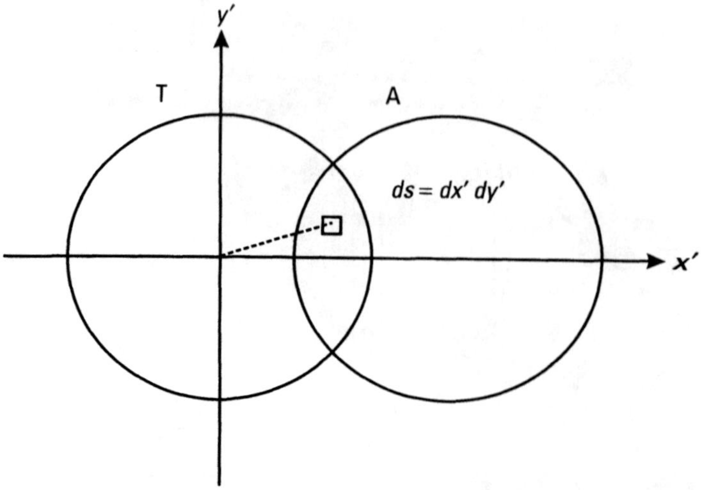

Figure 3.25 Transmission in the exit plane, SM fiber.

The function A($x'y'$), limited to S_1, corresponds to the incident amplitude in the plane $x'y'$. The function T($x'y'$) limited to S_2, corresponds to the amplitude transmission function of the fiber F_2. Therefore, the amplitude dA induced in F_2 for an elementary spectral width $d\lambda$ will be:

$$dA = A(x'y')T(x'y')d\lambda$$

When λ varies, the function A is translated in the plane $x'y'$.

If we assume that the dispersion is linear along x′:

$$\lambda - \lambda_0 = \alpha(x' - x'_0)$$

This is only an approximation in which the sine of the diffraction angle is approximated to the angle, for small angles.

The flux F is the squared modulus of the correlation of A with T:

$$\mathrm{F} = |\mathrm{K}\iint \mathrm{A}(x' - x'_0, y')\mathrm{T}(x'y')dx'dy'|^2$$

Let us assume that, in the Gaussian approximation, with a mode radius ω'_0 corresponding to the half width of the amplitude distribution A at 1/e:

$$\mathrm{A} = \exp - \left[\frac{r^1}{\omega'_0}\right]^2$$

With

$$r'^2 = x'^2 + y'^2$$

$$\mathrm{F} = \left|\mathrm{K}\iint_{+\infty}^{-\infty} \exp\left[\frac{x'^2 + y'^2}{\omega_0'^2}\right]\exp\left[\frac{(x' - x'_0)^2 + y'^2}{\omega_0'^2}\right]dx'dy'\right|^2$$

From which:

$$\mathrm{F} = \left|\mathrm{K}\int_{-\infty}^{+\infty} \exp\left[\frac{2y'^2}{\omega_0'^2}\right]dy'\int_{-\infty}^{+\infty} \exp\left[\frac{x'^2}{\omega_0'^2}\right]\exp\left[\frac{(x' - x'_0)^2}{\omega_0'^2}\right]dx'\right|^2$$

It can be seen that only the second integral depends on x_0'. Its direct calculation is relatively easy. But it is easier to consider this second integral as an autocorrelation function:

$$\left[\exp\left(\frac{x'^2}{\omega_0'^2}\right) \otimes \exp\left(\frac{x'^2}{\omega_0'^2}\right)\right]_{(x'_0)} = [g(x') \otimes g(x')]_{(x'_0)}$$

It is well known that, $f(x)$ being a function and $\tilde{f}(u)$ being its Fourier transform, the Fourier transform of the function $g(x) = f(x/a)$ is $\tilde{g}(u) = |a|\tilde{f}(a\,u)$ ("dilation" theorem), then:

$$g(x') = f(x' / \omega_0') = \exp-\left(\frac{x'}{\omega_0'^2}\right)^2$$

and

$$f(x') = e^{-x'^2}$$

As the Fourier transform of $\exp\left(-\pi x'^2\right)$ is $\tilde{f}(u) = \exp\left(-\pi u^2\right)$, the Fourier transform of $g(x')$ is:

$$\tilde{g}(u) = \sqrt{\pi}\,|\omega_0'|e^{-\pi^2 \omega_0'^2 u^2}$$

In the case of symmetrical functions, the correlation and the convolution are identical. Moreover, it is known that the Fourier transform of the auto-convolution product of a function $g(x')$ is the squared modulus of the Fourier transform of that function:

$$TF\left[g(x') \otimes g(x')\right] = |\tilde{g}(u)|^2$$

$$|\tilde{g}(u)|^2 = \pi\omega_0'^2 \exp\left(-2\pi^2 \omega_0'^2 u^2\right) = \pi\omega_0'^2 \exp-\left(\pi\sqrt{2}\omega_0' u\right)^2$$

Then, if we go back to the initial function through an inverse Fourier transform:

$$\left[g(x') \otimes g(x')\right] = TF - 1\left[|\tilde{g}(u)|^2\right]$$

and if we use again the dilation theorem on the inverse Fourier transform, we get:

$$\left[g(x') \otimes g(x')\right] = \sqrt{\pi}\frac{\omega_0'}{\sqrt{2}} \exp\left(-\frac{x'2}{2\omega_0'^2}\right)$$

or:

$$F = k \exp - \left(\frac{x'}{\omega'_0} \right)^2$$

Thus, the intensity transmission function F (λ) is identical to the function representing the amplitude distribution A (x') when the two functions are drawn on the same graph using abscissas $(\lambda - \lambda_0)/\alpha$ for F (λ) and $(x' - x'_0)$ for A (x') (Figure 3.26).

The total width at half maximum of F is:

$$2\omega'_0 \sqrt{\ln 2} = 1{,}6651\omega'_0$$

3.5.3.2 Devices with an SM Entrance Fiber and an Exit Slit or a Diode Array with Rectangular Pixels

Let us consider Figures 3.27 and 3.28. We assume that the coupling device is similar to that of Figure 3.23, but an exit slit is placed at $x' = 0$, $y' = 0$, the height of the slit being parallel to the grating grooves and y'.

We assume that the transmission P (x', y') is uniform and equal to 1 inside the slit, which is considered to have a height much larger than the fiber core diameter, corresponding to an intensity distribution I (x', y').

The elementary intensity dI transmitted through the slit for an elementary spectral width $d\lambda$ will be:

$$dI = I(x', y')T(x', y')d\lambda$$

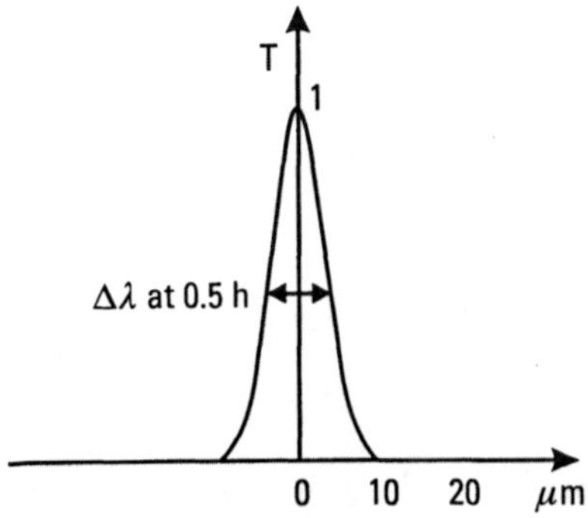

Figure 3.26 Transmission versus distance between the exit fiber center and the entrance fiber image center.

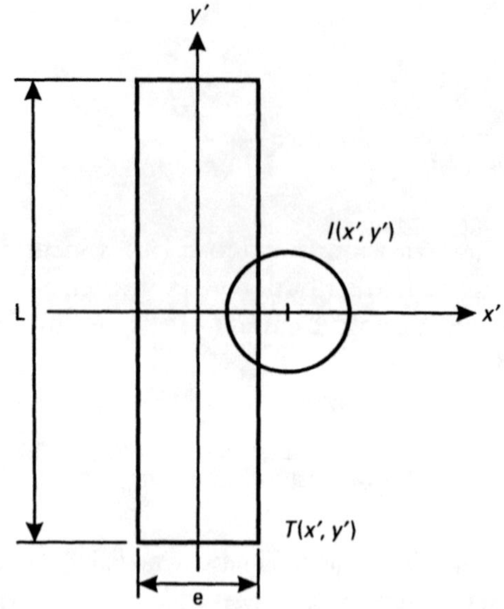

Figure 3.27 Exit slit and image of the entrance fiber in the exit plane.

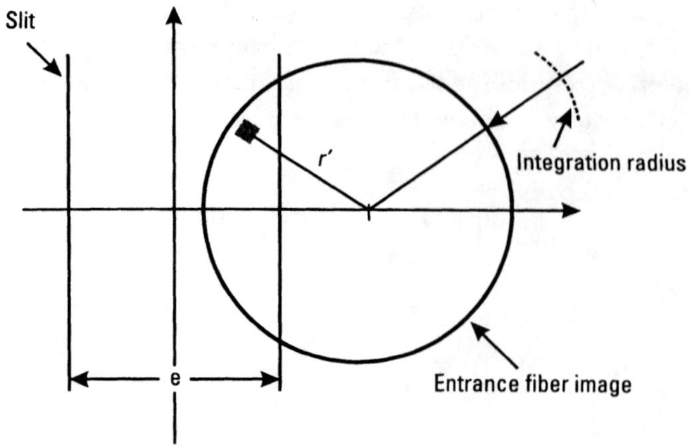

Figure 3.28 Notations in the exit plane.

As above, when λ varies, I is translated nearly linearly (same approximation as in Section 3.4.3.1). The transmitted flux is the modulus of the correlation of I with T:

$$F = K\iint I(x' - x'_0, y')T(x', y')dx'dy'$$

$$F = I \otimes T$$

The intensity distribution in the image (assumed perfect) of the SM fiber end-face is:

$$I = I_0 \exp - \frac{2r^2}{\omega_0'^2}$$

The double integral can be decomposed into simple integrals on r. Those integrals can be calculated using the Simpson method. We used only 11 elements, which correspond to a precision better than the measurement precision coming from experimental uncertainties. The maximum integration radius is 20 μm. We used slit widths of 5, 10, 12, 14, and 20 μm. Only for e = 5 μm, is the integration method slightly different, and hence more precise. The results are given on the curves in Figure 3.29 for e = 5, 10, 12, and 20 μm, Figure 3.30, and Figure 3.31 for e = 10, 12, and 14 μm compared to the results obtained with devices using input and output SM fibers.

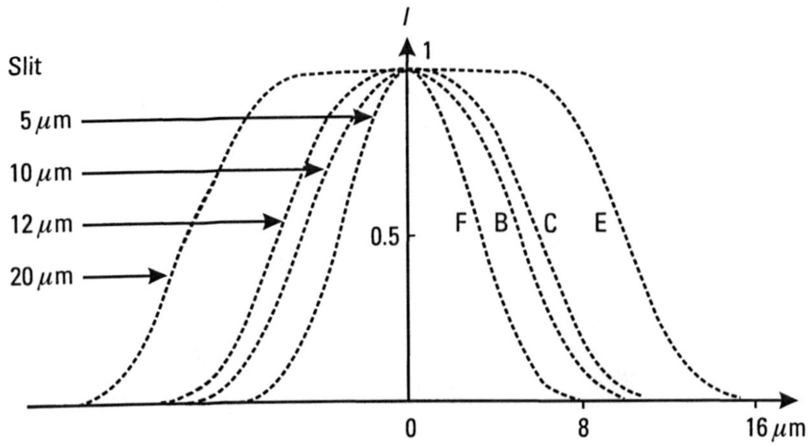

Figure 3.29 Intensity transmission functions. SM fiber to exit slit. SM fiber, mode radius $\omega_0 = 4{,}573$ μm in slit F, slit width: e.

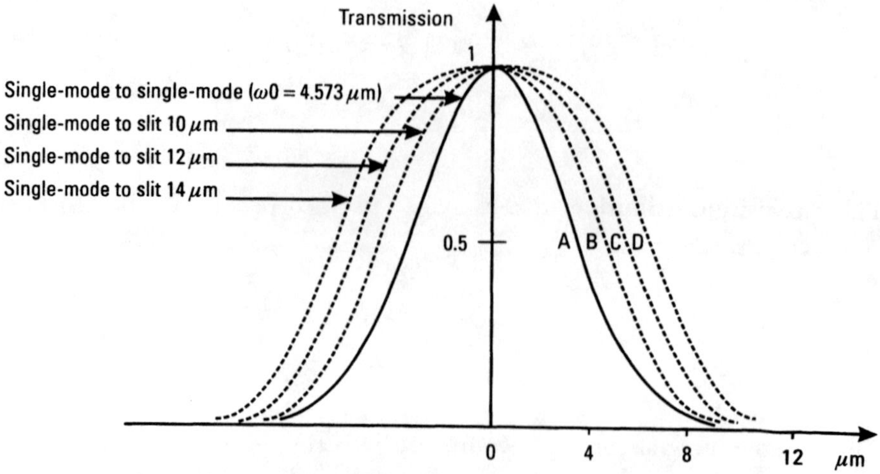

Figure 3.30 Intensity transmission functions. SM fiber in, exit slit. SM fiber, mode radius $\omega_0 = 4{,}573\,\mu$m in an identical SM fiber and in slits with width e.

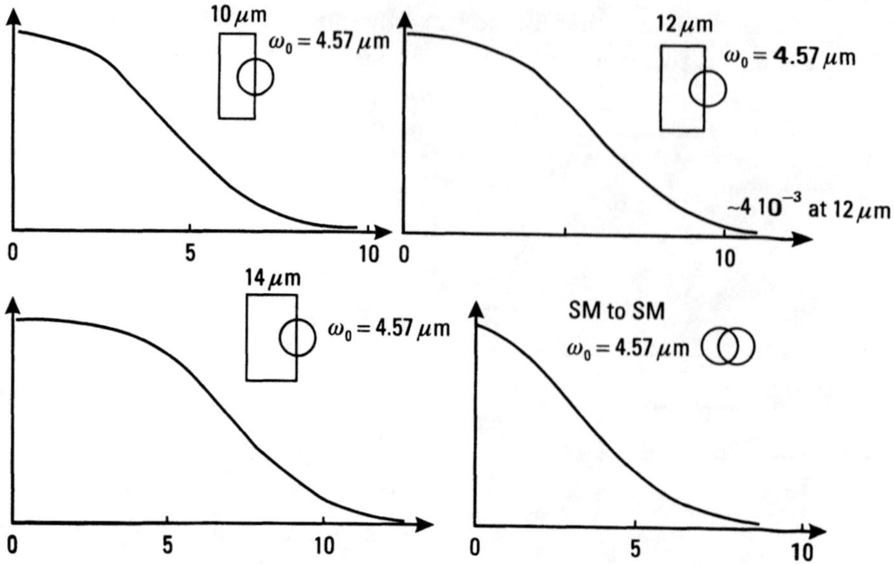

Intensity in the entrance fiber: $I = I0 \exp\left|\frac{-2r^2}{\omega_0^2}\right|$

Figure 3.31 Comparison of intensity transmission function of an SM fiber with a mode radius $\omega_0 = 4{,}573\,\mu$m, through another SM fiber with a mode radius ω_0, or through exit slits with different widths: 10 μm, 12 μm, 14 μm, 20 μm.

3.5.3.3 Devices with an SM Entrance Fiber and Multimode Step Index Exit Fibers

We consider that a multimode fiber with a numerical aperture that is always larger than the numerical aperture of the SM fiber collects all the available intensity in its core having a radius R_1 (Figure 3.32). The integration is limited to a domain of radius R_2 much larger than ω_0' but with $R_2 < R_1$.

Let us consider an elementary surface $ds = 2\alpha\ rdr$. We must identify four cases:

1) $x_0' > R_1 + R_2 \quad F = 0$

2) $R_1 < x_0' < R_1 + R_2$

$$F = \int_{R_2}^{x'-R_1} 2r \exp\left(\frac{-2r^2}{\omega_0'^2}\right) Arcos\left(\frac{r^2 + x_0'^2 - R_1^2}{2r\,x_0'}\right) dr$$

3) $R_1 - R_2 < x_0' \le R_1$

$$F = \int_{R_2}^{R_1 - x_0'} 2r \exp\left(\frac{-2r^2}{\omega_0'^2}\right) Arcos\left(\frac{r^2 + x_0'^2 - R_1^2}{2r\,x_0'}\right) dr$$

$$+\pi \frac{\omega_0'^2}{2}\left(1 - \exp\left(-\frac{2(R_1 + x_0')^2}{\omega_0'^2}\right)\right)$$

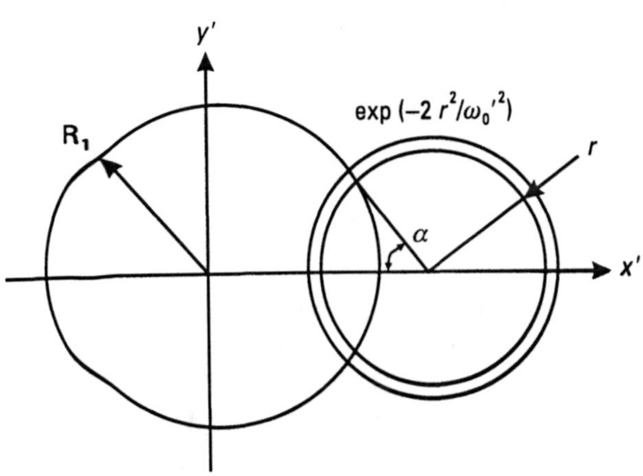

Figure 3.32 Multimode fiber core intercepting an SM fiber intensity distribution.

4) $x_0' < R_1 - R_2$

$$F = \pi \frac{\omega_0'^2}{2}\left(1 - \exp\left(-\frac{2R_2^2}{\omega_0'^2}\right)\right)$$

We calculated these functions for the different cases (see Figure 3.33).

3.5.3.4 Devices with an SM Entrance Fiber and Multimode Graded Index Exit Fibers

A similar mathematical development leads to the following transmission curves (see Figure 3.34 and Table 3.4):

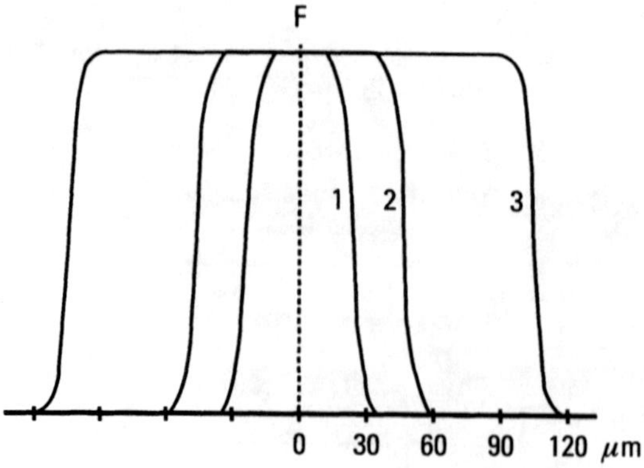

Figure 3.33 Entrance SM fiber. Exit step index fiber. (Curve 1: $R_1/\omega'_0 = 9.09$, Curve 2: $R_1/\omega'_0 = 18.18$, Curve 3: $R_1/\omega'_0 = 36.36$.)

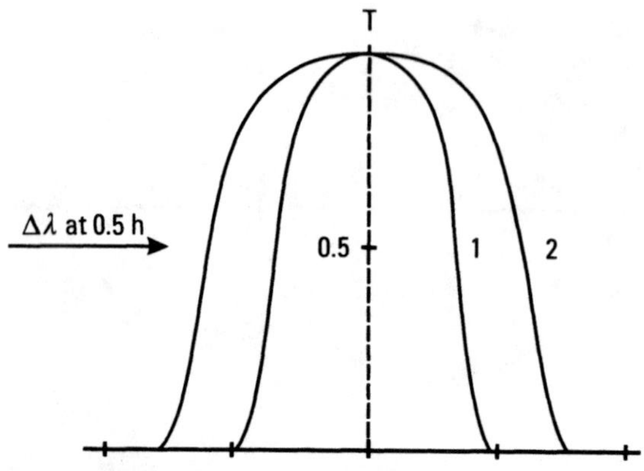

Figure 3.34 SM entrance fiber and graded-index exit fiber.

Table 3.4
Parameters of the Curves in Figure 3.34

Curve	1	2
Entrance core Diameter in μm	8	8
Exit core Diameter in μm	50	85

3.5.3.5 Coupling in a Device with Small Aberration

In such a case, from a point located at the center of the entrance fiber core, the device gives an image spot that is well focused within a few micrometers. Thus, the coupling losses are negligible in the single-mode to multimode case (usual demultiplexer: SM → 50 × 125 μm, for example) and the F function is not modified much.

This is not true when coupling between SM fibers (multiplexer) is considered. In the entrance fiber image spot, the amplitude distribution A′ is modified by the aberrations. The new distribution A′ will be the convolution product of A by the instrumental response P of the system to an elementary source:

$$A' = [A * P]_{(x_0')}$$

We will obtain:

$$F = K\left|[A * P \otimes A]_{(x_0')}\right|^2$$

The calculation is generally easier when done in the exit pupil plane. It can be demonstrated that the coupling losses remain smaller than 1 dB as long as the sum of the different aberrations measured in the exit pupil remains smaller than $\lambda/3$ [49].

3.5.4 Grating Micro-Optic Devices

Generally, grating multiplexers or demultiplexers consist of three main parts: entrance and exit elements (fiber array or transmission-line fiber and emitters or receivers), focusing optics, and dispersive grating. The grating is often a plane grating [50–58].

For instance, the Finke, et al. setup is very simple in principle. The fiber array extremity is at the object focus of a ball lens, the grating being located at the image focus. The system is afocal, with magnification 1, so that all ray angles from and to the fibers are identical, as originally proposed for the configuration of Figure 3.36. The authors claim 1.2- to 1.7-dB demultiplexer losses with four to six channels. Demultiplexers with up to 10 channels with multimode fibers were designed around such a configuration.

In 1977, another configuration was proposed by Tomlinson (Figure 3.35) [52, 55], and it is still often used. It consists of a graded index lens placed in front of a plane grating prism. In Europe, Mannschke obtained interesting results with a five-channel multimode fiber demultiplexer; losses were 0.9 to 2 dB [59–60].

In 1980, we published the configuration of Figure 3.36: A fiber array is placed in front of a slit, photo etched on a plane reflection grating, perpendicular to the grooves. A concave parabolic mirror transforms the diverging beam coming from any fiber into a parallel beam; this beam coming to the grating is angularly dispersed back to the concave mirror and is imaged on the fiber array extremity in a position that depends on its wavelength. This configuration is aplanetic, afocal, and has a magnification of 1. Thus, all angles from and to the fibers are identical, and we obtain the best conditions for a high coupling efficiency. The achromatism is perfect and aberrations are almost nil when the mirror is parabolic. Indeed, a spherical mirror is often used because the aberrations remain very small. In each case, with monochromatic sources centered on the multiplexer transmission bands, the losses remain smaller than 2 dB and may reach 0.5 dB, as in the

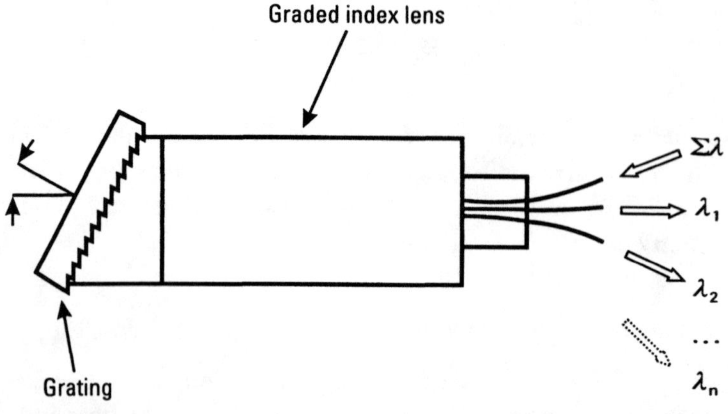

Figure 3.35 Multiplexer with a plane grating and a graded-index lens. (*After:* Tomlinson [52–55].)

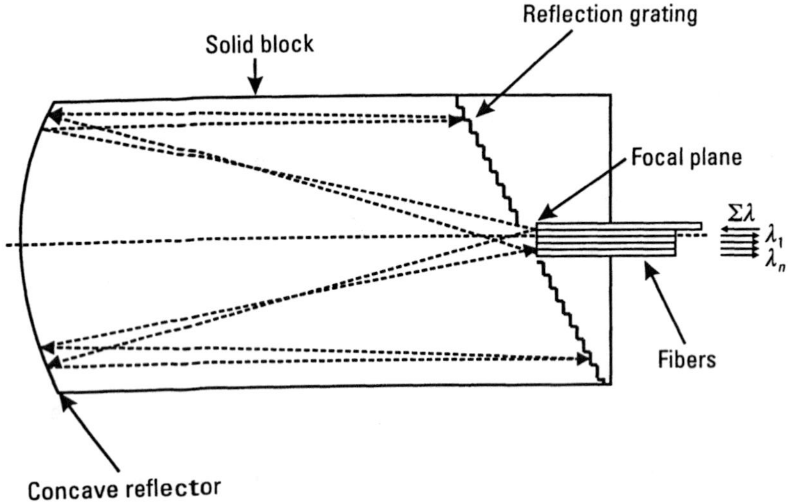

Figure 3.36 Multiplexer using SM or multimode fibers. (*After:* J. P. Laude [57].)

1.54/1.56 μm single-mode device of [61]. In many cases, for cost saving, a spherical mirror with a Mangin-type aberration correction is preferred here over the parabolic mirror solution [62]. Using two materials with identical chromatic dispersion, the geometrical aberration can be corrected without introducing chromatism.

The use of dedicated CGs [57, 63–65], simplifies the device. For example, in the early Harada, et al. configuration (Figure 3.37) in which 2.6-dB losses are reached on a four-channel multiplexer at 821.5, 841.3, 860.9, and 871 nm with 60/125 μm multimode fibers. However, it is impossible to retain aberration-free focusing over a large spectral range with such

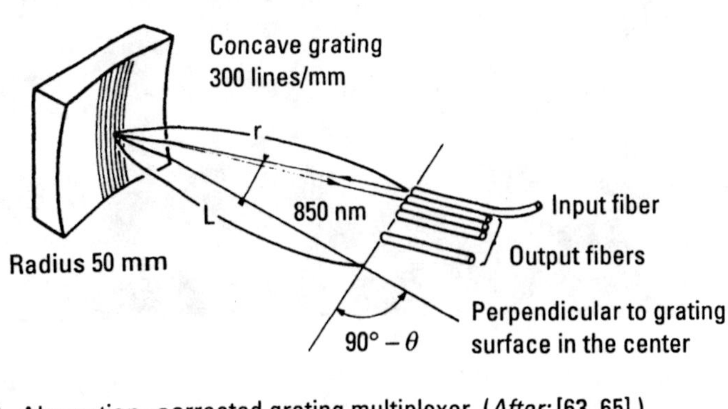

Figure 3.37 Aberration-corrected grating multiplexer. (*After:* [63, 65].)

configurations. The devices from [66, 67] are also interesting multiplexer examples from the early 1980s.

The multiplexer is always more difficult to manufacture than the demultiplexer because the core diameters are small and the highest optical quality is essential. This is particularly true when all SM fiber arrays are used. For example, if a $\lambda/10$ defect of spherical aberration, coma, and astigmatism is allowed, the cumulative losses are 1 dB [49, 68]. Moreover, in order to obtain a channel spectral width large enough, as compared with channel spacing, the geometrical distance between fiber cores must be small. The pass-bands are practically adjacent for a distance between fibers of 22 μm when a 11 μm diameter core is used (Figure 3.38). Fibers with a small cladding are necessary. This can be obtained by chemical etching of larger core fibers. As early as 1984, Hegarty, et al. [69] manufactured a nine-channel SM multiplexer, with a distance between fibers of 36 μm. The width of each channel was 0.2 nm at a wavelength about 1.5 μm, with 1.5 dB losses.

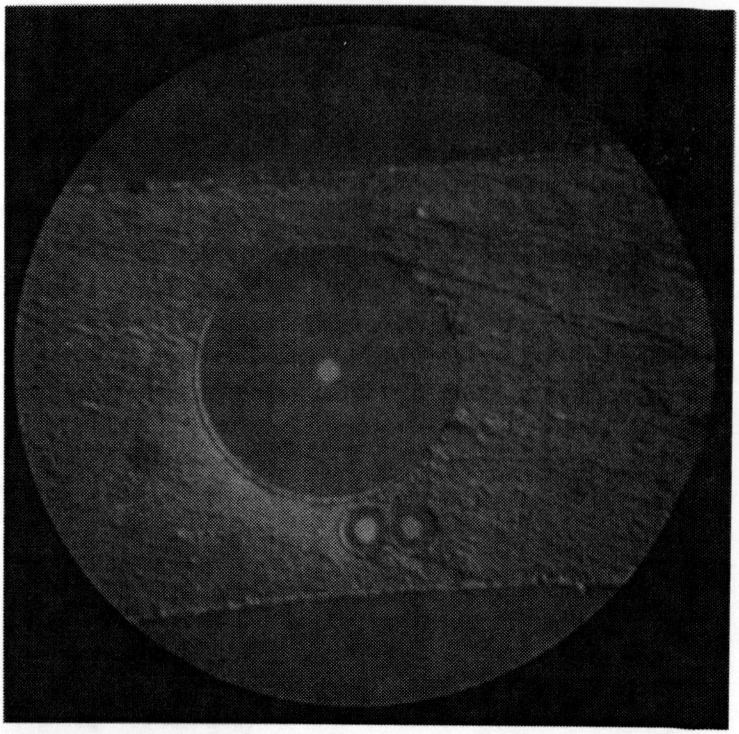

Figure 3.38 SM fibers at the focal plane: One 11/125 μm and two 11/20 μm, fibers spaced at 22 μm. (*Source:* HighWave Optical Technologies.)

In the ideal case, (zero aberration device, all SM demultiplexer, the adjustments being assumed perfect) and in the Gaussian approximation of the single-mode fiber electric-field description, the ratio R_w between the width at half-maximum $d\lambda_{Fwhm}$ of the transmission functions and the spectral distance between channels $\Delta\lambda$ is

$$R_w = \frac{d\lambda_{Fwhm}}{\Delta\lambda} = \frac{1.66\omega_0'}{\Delta x'}$$

($\Delta x'$ is the distance between fibers at the focus and ω_0' is the mode radius).

On practical stigmatic DWDMs the agreement between this theoretical value and the experimental results is quite good (within a few percentages) [70, 71].

In order to obtain optimized spectral pass-bands without requiring a small distance between fiber cores (as is necessary in simple devices), different solutions have been proposed, such as a spectral recombination within each channel via an intermediate spectrum image, through microprisms or through a microlens array, as in [72], in which $R_w = 0.7$ is obtained with up to 32 channels. However, the manufacturing cost and the additive losses of such devices remain a drawback.

3.5.5 Thermal Drift of Grating Micro-Optic Devices

Thermal drift depends on the thermal expansion coefficients (of the different materials and on the index variation *dn*/*dt*. For monoblock grating WDM, it has been shown [73] that the wavelength shift is given by:

$$\Delta\lambda/\lambda = (\varepsilon + 1/n\, dn/dt)\Delta t$$

$\Delta\lambda = 12$ pm/°C with silica at 1,550 nm.

For some devices using a grating in air, the thermal drift can be reduced to about 1 pm/°C.

For monoblock grating DWDMs, the thermal expansion of the grating in silica can be compensated by a negative *dn*/*dt* of the optical block (Figure 3.39). Let us give two typical results from the HighWave laboratory:

- On a 35 channel, 25-GHz-spacing athermal DWDM, the passive drift was reduced down to 0.4 pm/°C. (Figure 3.40).

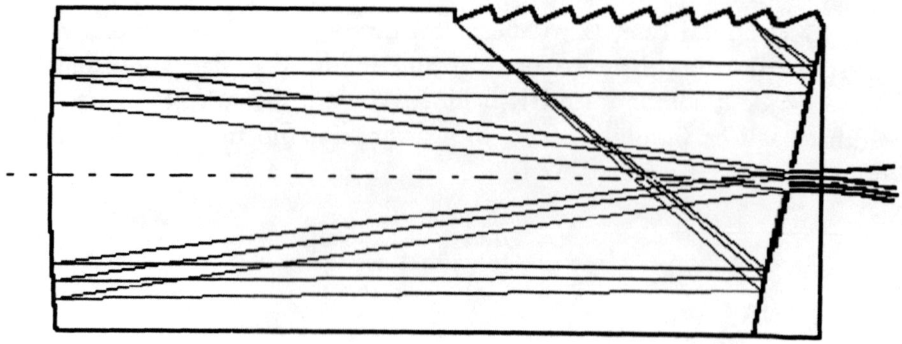

Figure 3.39 High Littrow angle-grating configuration for very dense WDM.

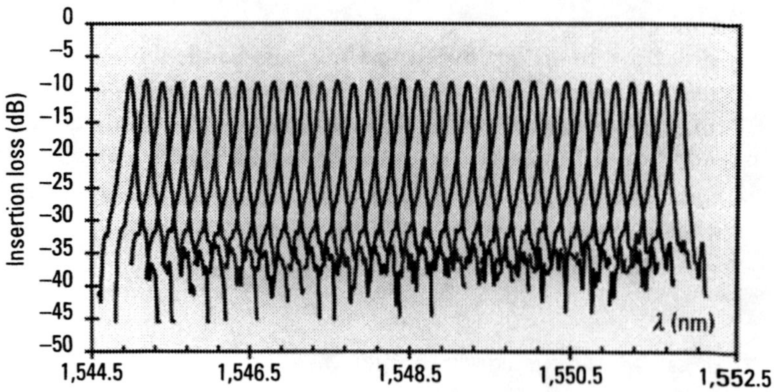

Figure 3.40 OG athermal DWDM transmission functions: 35 channels, 25-GHz spacing. (*Source:* J. P. Laude [74].)

- On a 160 channel, 25-GHz-spacing athermal DWDM, built from two individual interleaved 80 channel, 50-GHz-spacing components shifted by 25 GHz, the passive drift was reduced to 0.35 pm/°C. (Figure 3.41 and Table 3.5).

3.6 Cascaded Mach-Zehnder Interferometers

Mach-Zehnder (MZ) devices are based on interference between two light beams with optical path length difference $n\ \Delta L$ between both arms. The interferometer consists of two fibers with different lengths coupled together at each end with a 3-dB coupler, or can be preferably built with two

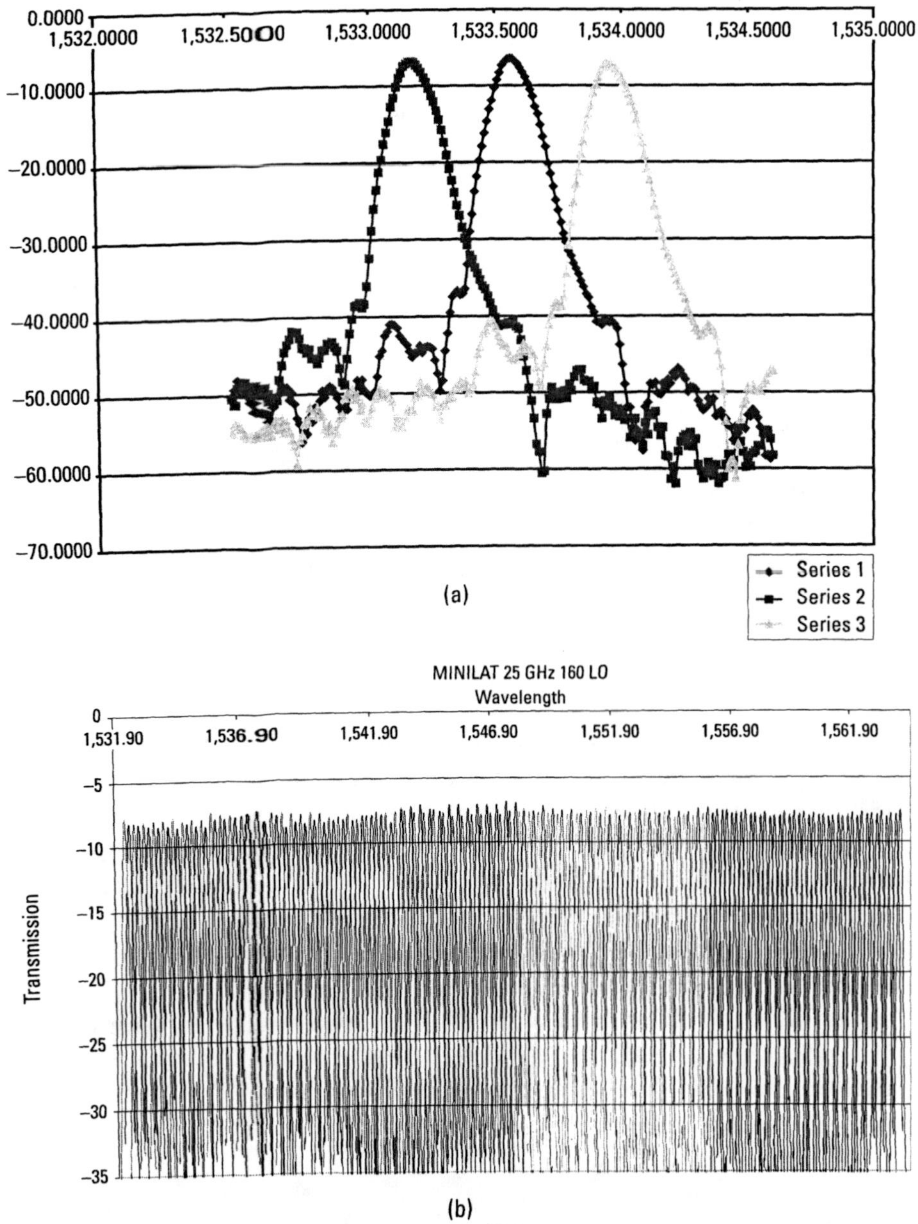

Figure 3.41 ITU 25-GHz spacing, 160-channel VD-WDM with two interleaved DWDM units: (a) details of three adjacent channels of one DWDM unit, and (b) 160 channels of two interleaved DWDM units. Left side details three adjacent channels of one of the interleaved units. (*Source:* J. P. Laude [75].)

Table 3.5
Monoblock Grating DWDMs: Typical Data

Parameter	Data		Unit	Notes
Number of channels	80	160	—	ITU 195.600 to 191.625
Wavelength spacing	50	25	GHz	≈ 0.4 nm/≈ 0.2 nm
Insertion loss	5	8.5	dB	50-GHz/25-GHz spacing
3-dB passband *	0.12	0.12	nm	*Up 0.18 with added loss
Crosstalk, adjacent channels	<–33	<–35	dB	50-GHz/25-GHz spacing
Polarization dependent shift	<< 0.1*	GHz	*Not measurable	
Polarization dependent loss	0.7 to 1 **	dB	**Feasibility < 0.5 dB	
Passive thermal drift	< 1	pm/°C	Down 0.35 pm/°C obtained	

(*Source:* HighWave Optical Technologies.)

waveguides and two couplers integrated on an optical chip. The wavelength transmission functions of such devices are periodic, 90 degrees out of phase with one another on the two branches of the exit coupler.

In the first two chips (Figure 3.42), when the length difference between the two paths is ΔL, the index is *n*, the transmission function is: $T(\lambda) = \cos^2(\pi n \Delta L/\lambda)$.

We get a periodic filter in frequency ν: $T(\nu) = \cos^2(\pi\nu\tau)$ where τ is the time difference between the 2 arms: $n\Delta L/c$.

In order to get narrow bandpasses, MZ interferometers with optical paths $n\Delta L$, $2n\Delta L$, $4n\Delta L$, ... $Mn\Delta L$ can be cascaded [76]. For M stages we can isolate one channel among N channels with $N = 2^M - 1$ equidistant channels, with a spacing $\Delta\nu$.

The transmission function is:

$$T(f) = \left[\frac{\sin(\pi\nu n\Delta L\Delta\nu)}{N\sin(N\pi\nu n\Delta L\Delta\nu)}\right]^2$$

As the number of stages increases the function approximates the transmission function of an FP filter.

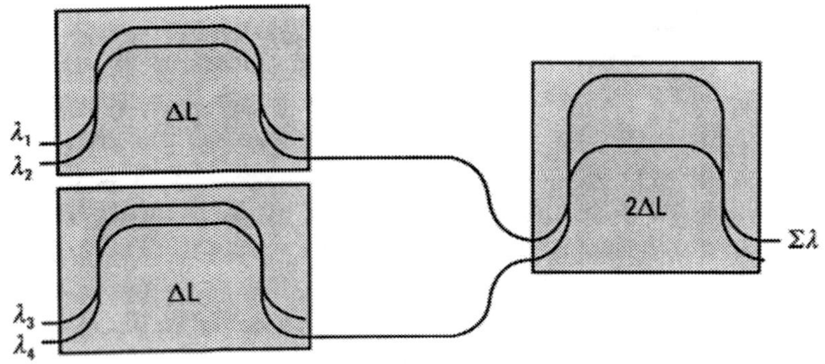

Figure 3.42 Cascaded MZ interferometers.

For simultaneously demultiplexing all N channels, the demultiplexer would require ($2^M - 1$) MZ units. Therefore, the device is generally used with few channels only, or as a tunable filter.

Following are a few examples of practical design performances. Takato, et al. designed up to 16-channel WDM using cascaded integrated optics (IO) MZ interferometers [77]. The devices are said to have insertion losses of approximately 0.5 dB in the WDM region (a few THz spacing) and 2 to 5 dB losses in the FDM region (a few gigahertz spacing). The same NTT team [78, 79] designed a 128-channel DWDM device with seven-cascaded MZ, on a 5- by 6-cm silica silicon substrate with 6.7 loss and −1-dB crosstalk.

In the ACTS European research COBNET, IBM Zurich laboratory designed an optical add-drop subsystem with cascaded MZ elements on a high-index-contrast silica on silicon technology based on PECVD-deposited siliconoxynitride (SiON) allowing a small bending radius (1.5 mm). The FWHM was 1.1 and 1.6 nm (add and drop, respectively) for a 75- mm-long, 4-channel device working at 1.6-nm spacing. The fiber to fiber losses were 5.4 to 6.8 dB, 6.0 to 7.0 dB, and 8.2 to 9.0 dB, for an input to drop path, an input to through path, and an add to through path, respectively. The isolation was 28 to 33 dB and 22 to 33 dB, for drop channel and transit channel, respectively.

3.7 Other Devices: FBG/MZ Interferometer Devices

The idea of placing FBG inside the two arms of a Mach-Zehnder interferometer (MZI) to make a wavelength-selective coupler was first proposed in the 1980s by a team from the Canadian Communication Research Center.

Since then, many other groups designed such components in both fiber and IO form [80]. In principle, these devices have less insertion loss and potential cost than FBG using circulators for wavelength extraction. A 50% coupler splits the incoming light into the two arms of the interferometer (Figure 3.43). In each arm an identical Bragg grating reflects one wavelength of the incoming channels while passing all the other channels. Of course, Bragg gratings with different periods can reflect several wavelengths. Another coupler combines the transmitted channels. A phase shift between the two reflected signals is such that the reflected signals are out of phase, canceled, on the input arm of the first coupler, and added in phase in the second arm. So the reflected signals appear on its "drop" port. The transmitted signals are added in phase in the "through" port of the second coupler.

3.8 Methods for Broadening and Flattening the Spectral Shape of the Transmission Channels of Grating WDM

3.8.1 Introduction

Let us consider the case of grating WDM using integrated optics (IOG): AWGs, and CG solutions, or three-dimensional optics components (bulk optics or aerial configurations with a classical or holographic diffraction grating). In these grating devices the spectral response is theoretically a Gaussian function for low-loss coupling devices [73] imposing some constraints on wavelength control. Different solutions were proposed for flattening the spectral response of AWG: using a multimode interference coupler in the

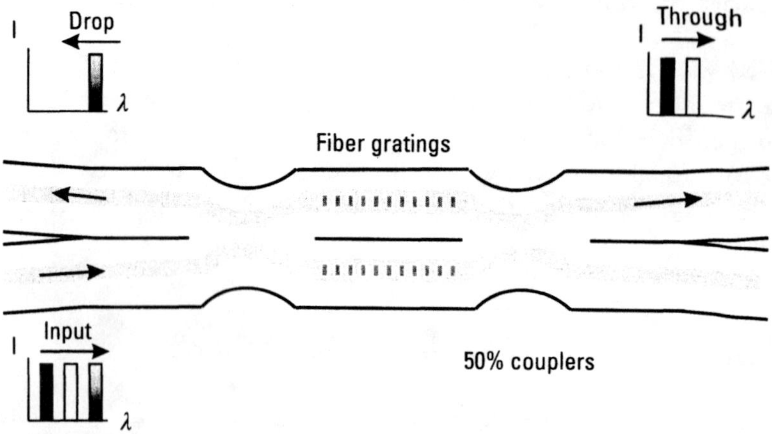

Figure 3.43 FBG/MZI.

input waveguide [81], using two focal points in the input and output star couplers [82], a double-phase array [83], or a phase filter on the grating [84]. The fundamental limits and conditions on the phase filters that must be realized to approach rectangular passbands on AWG were derived [85]. All these methods for broadening and flattening the passband that can be applied to AWG as well as to IOG, bulk or aerial optics grating devices for flattening the spectral response. They can be generalized to all grating solutions [86], and are not only applicable to gratings but also to any WDM component with angular dispersion of wavelengths.

3.8.2 Principle

In the most general case, the WDM multi/demultipler or router is the device in Figure 3.44. We have at least one fiber F1 (the line) located at the object plane (xy), one fiber F2 (one of the fibers supporting the different wavelengths) at the image plane $(x'y')$, a coupling device with a grating or a wavelength disperser R located at the pupil, $(\alpha'\beta')$. Our calculations are made for three-dimensional devices but lead to similar conclusions for AWG and IOG. Let us calculate the transmission function F from F1 towards F2. Let us consider a fiber F1 with a polished end-face in the plane xy perpendicular to the direction zz' axis of the fiber. The end-face of F1 is imaged on the plane $(x'y')$, which contains the polished end-face of fiber F2, after an angular dispersion on R (generally a classical grating or a phase array grating). We will calculate the energy transmitted through F2 as a function of the wavelength of the light delivered by L2. We assume hereafter that the magnification is 1 and that the coupling is afocal. As a matter of fact, this condition is necessary to avoid mode-coupling loss between identical fibers.

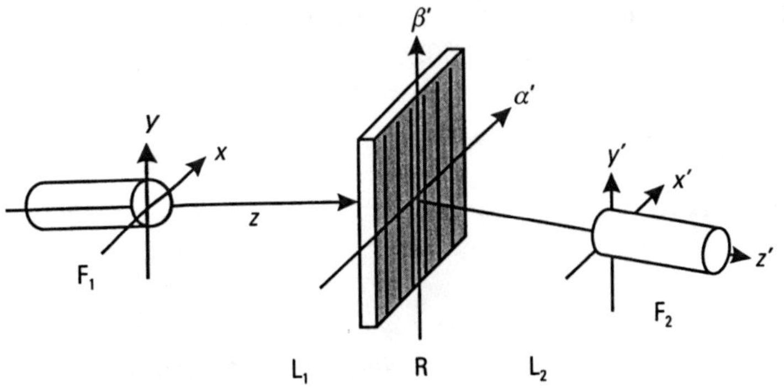

Figure 3.44 Notations: (xy) entrance plane, $(\alpha'\beta')$ exit pupil plane, $(x'y')$ exit plane. F1 and F2 in and out fibers.

Let A $(x\ y)$ be the incident amplitude in the $x\ y$ plane, and T $(x\ y)$ be the amplitude transmission function of the exit fiber F2. Therefore, the amplitude dA induced in F2 for an elementary spectral width $d\lambda$ will be: $dA = A(x'y')T(x'y')d\lambda$.

When λ varies, the function A $(x'y')$ is translated in the plane $x'y'$. If we assume that the dispersion is linear along x: $\lambda - \lambda_0 = \alpha(x' - x'_0)$. This is only an approximation in which, for small angles, the sine of the diffraction angle is approximated to the angle. The flux F will be the squared modulus of the correlation of A with T, with identical fibers:

$$F = \left| K[A \otimes A]_{x'_0} \right|^2$$

with SM fibers, A $(x'y')$ and T $(x'y')$ are Gaussian functions in a "perfect" optical system with no limitation at the pupil, then F is a Gaussian function of x', and then also of λ [78].

Now we can introduce a spatial filtering in the pupil plane [86]. It can be, for instance, a variation in the optical path. Let us suppose that the filtering function is $S = \text{sinc}(\Pi\, a\, \alpha')$. Its Fourier transform in the image plane will be: P (x') = 1/a Rect (x'/a). The new amplitude distribution in the image is the convolution product A∗P so we get a transmission:

$$F = \left| K[A * P \otimes A]_{(x'_0)} \right|^2$$

F becomes a flat-top function, with the width of the top depending on a. There are no secondary lobes.

Realistically, it is very difficult to design such a filter, but the effect of S on the Gaussian field G in the pupil can be approximated. We can manufacture a filter with one or several Π phase difference zones in the plane $(\alpha'\beta')$, approximately simulating the effect of the first lobes of the sinc $(\Pi\, a\, \alpha')$ function (Figure 3.45) when combined with the Gaussian field in $(\alpha'\beta')$.

Preliminary good results were obtained with a pure dephasing mask with a phase transition variation along β'. The first negative lobe being located in the foot of the electrical field incident in the $(\alpha'\beta')$ plane, the amplitude of the product can be adjusted to the mean value of the first negative lobe of sinc $(\Pi\, a\, \alpha')$. However, the steep variation of phase in β' produces small secondary lobes in the diffraction function that gives interchannel crosstalk, typically –25 dB at 0.4 nm (50 GHz) and –42 dB at

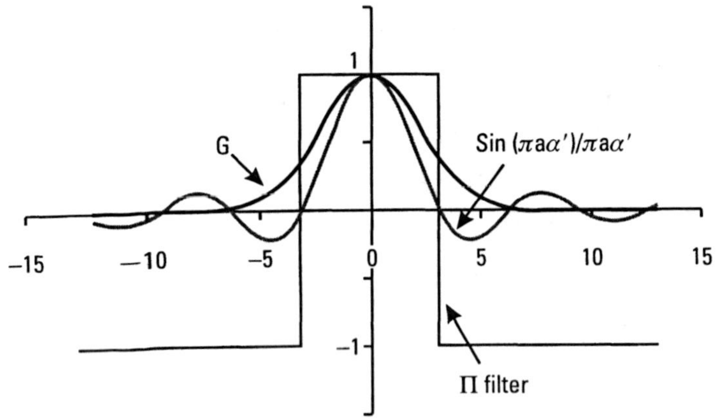

Figure 3.45 Filtering functions in the pupil plane: Sinc ($\Pi a\alpha'$) would be the ideal filtering; however, a Π dephasing filter reproducing the sign of the ideal function combined with incident field G already gives interesting results.

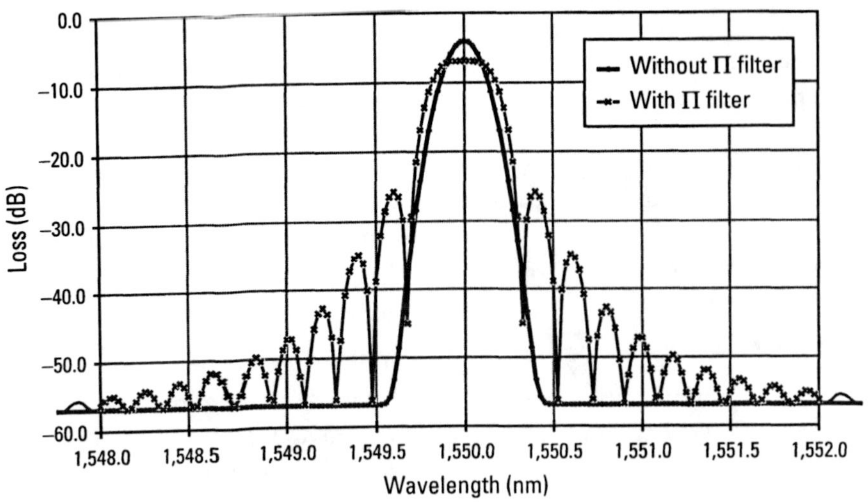

Figure 3.46 Theoretical wavelength response with 100-GHz channel spacing with and without Π filter.

0.8 nm (100 GHz). The theoretical curves given in Figure 3.46 give a comparison of transmission functions of a 100–GHz channel-spacing grating WDM with a theoretically nonlimiting pupil, with or without the filtering. It will be possible to modify the phase transitions further in order to get better results. (All these results are given for identical input and output SM fibers. Of course, they can be dissimilar: For instance, it is always possible to manufacture a demultiplexer with a SM fiber input and MM fiber outputs. In such a case, it is much easier to get very flat passbands with or without a phase filter.)

3.8.3 Experimental Example

Let us take a 3D-WDM all-single-mode fiber device using plane reflection grating in Littrow configuration (Stimax [61]) with 32 channels at 100-GHz spacing with a Π phase mask with phase transitions made with a dielectric coating, corresponding to the calculations above. On this real configuration we also have to take into account the finite width of the pupil and the masking effect of the lithographed window typical of this configuration. In Figure 3.47 we compare the experimental results of devices with and without the Π spatial frequency mask. Without filtering we get Gaussian functions and bandwidths of 0.11 and 0.19 nm at −1 and −3 dB, respectively, to be compared with a much flatter transmission with filtering and bandwidths at −1 and −3 dB of 0.26 and 0.35 nm, respectively. The agreement between theoretical and experimental results is within a few percentages. The main differences between theory and experiment come from approximating the electrical field amplitude as a perfect Gaussian field in our calculations, and from the effect of the finite width of the pupil. Calculations taking into account the real pupil give much better agreement at larger distances from the maximum: −49 and −43 dB at 1.6 nm (200 GHz), respectively, for theory and experiment.

3.8.4 Conclusion

In this chapter we described a method for flattening the transmission channels of WDM components. This method gives interesting experimental results on bulk optic devices but can also be applied to many other configurations, such as IOG, in which the spatial phase modification can be applied on the grating, or anywhere far enough from the spectral plane. Or it can be applied on AWG, for instance, by modification of the length of the discrete waveguides following a phase law in sinc $(\Pi \, a \, \alpha')$ that can be approximated easily.

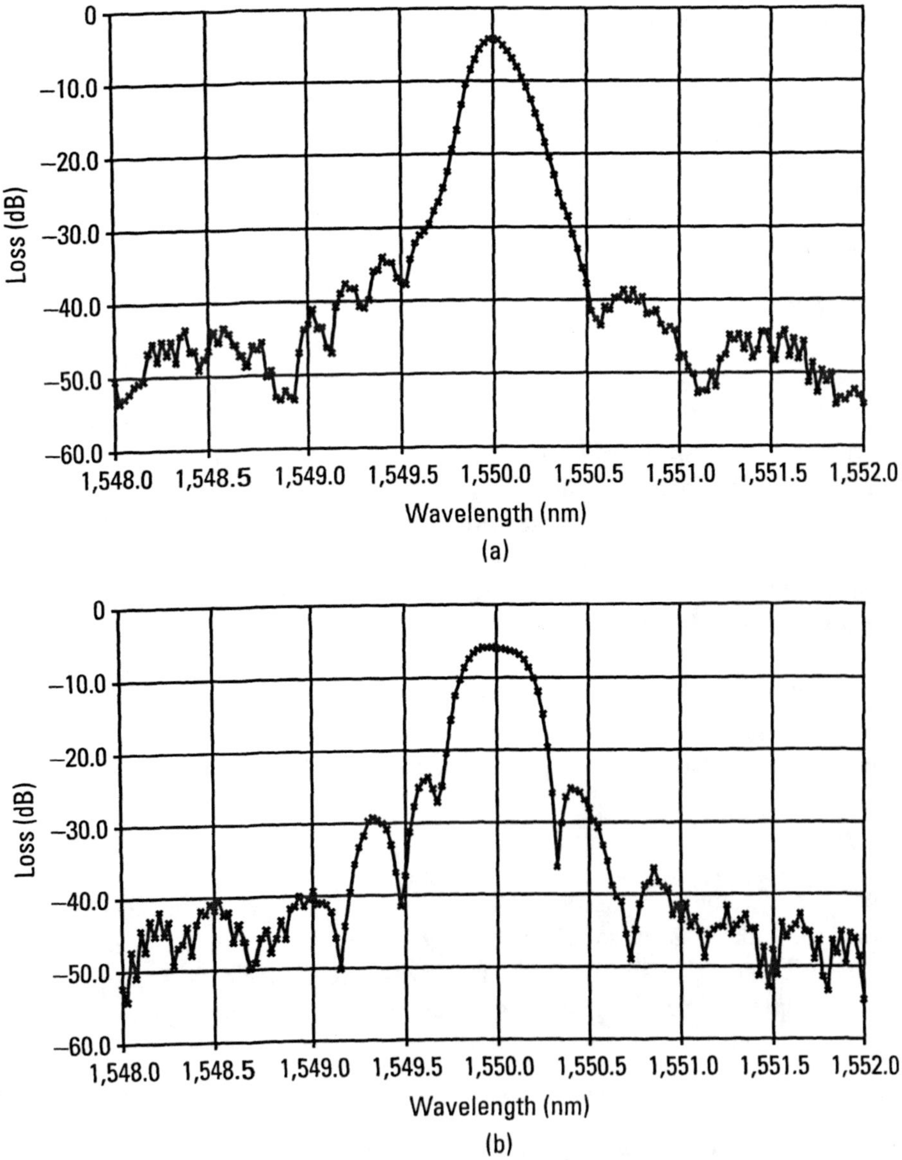

Figure 3.47 Experimental results with 100-GHz channel spacing: (a) without a filter, (b) with a Π filter, as in Figure 3.45.

3.9 Comparison of the Different Solutions

3.9.1 Some Remarks

3D optics filters (OF)

OFs can be made virtually temperature insensitive (< 2 pm/°C) but have higher and higher losses and complexity increasing with the number of channels.

Integrated optics AWG

AWG patents are awarded to large telecommunication companies. AWGs are claimed to be excellent for very close spacing. However, they suffer from larger losses, low free-spectral range, and relatively high crosstalk which limits the practical number of channels and restricts the AWG devices to monodirectional use. Their intrinsic thermal drift is 12 pm/°C for AWG on silica, and an order of magnitude more with AWG on InP. Some scientific papers show how to compensate this, lowering it to about 2 pm/°C, but these solutions could be considered rather academic. The practical solution for AWG remains the use of Peltier or other thermal stabilization.

Integrated optics concave gratings (CG)

To our knowledge there are no commercial devices available. These devices seem to have relatively large polarization effects, large crosstalk, and large losses.

FBG

FBG is generally preferred when just a few channels have to be extracted along the link. A good square wavelength filter shape can be created. The main problem is that FBG must be used with circulators or MZ couplers (as many couplers as wavelengths are to be extracted or inserted separately) adding losses and costs.

3D OG

3D OGs have tremendous capability in a number of channels (160 and more with bidirectionality commercially available). Small spacing such as 25 GHz is commercially available and 5-GHz feasibility was demonstrated in 1997. In the ACTS WOTAN research contract a 6.25-GHz spacing, 804-wavelength router was calculated. Athermal glass components down 25-GHz spacing are manufactured with uncontrolled drifts as low as 0.35 pm/°C using commonly available glasses.

These solutions are based on the traditional and mature optics technology used to manufacture millions of diffraction-limited reliable optics at low costs. The unique three-dimensional nature of OG is exploited for advanced features requested by network designers such as add/drops, cross connects, and routing devices.

3.9.2 Device Polarization Sensitivity

Critical to all devices is polarization sensitivity. It must be small, since transmission lines generally do not maintain the polarization state. In IOG devices, birefringence causes wavelength shifts: Even for the best new germanium-doped waveguides from NTT, the birefringence remains about 10^{-4}. However, polarization compensation can be designed.

Some three-dimensional optics devices show very low birefringence (e.g., 5×10^{-7} in pure silica). Consequently, the polarization mode dispersion (PMD) can be very small and there is no wavelength shift of the channel center with polarization. But the grating can induce polarization dispersion loss (PDL). PDL is lower than 0.5 dB with adequate special profile gratings. Moreover, PDL can be canceled by different means, such as a second pass on the grating after 90 degree rotation of the electric field.

3.9.3 Conclusion

Each solution has an application domain. For a low number of channels, filter and Bragg gratings have some advantages. AWGs are useful for monodirectional links. In the long run, if the AWGs can be integrated with active optoelectronics devices, they could become one of the most competitive solutions for local networks. Three-dimensional optics gratings can be used for high quality bidirectional links. For monodirectional or bidirectional links, three-dimensional optics gratings have virtually unlimited numbers of channels and provide the lowest crosstalk for the highest number of channels. For low-cost mass production the problems of fiber handling, which are the same with any solution, remain to be solved. In Table 3.6, typical performances of DWDM components available in 2001 are given.

Table 3.6
Typical DWDM Performances in 2001

Solutions	Number of Channels	Channel Spacing (nm)	Typical Loss dB	Xtalk dB	Free Spectral Range nm	Polarization Sensitivity % of $\Delta\lambda$	Intrinsic Temperature Drift nm/°C (typical)
AWG (NEC, NTT Electronics Corp., Lucent, Nortel, Pirelli, Corning, Hitachi, PIRI…)	Available up to 80	Down to 0.4 (But lower spacing already tested at research stage)	6 to 10	–25	Limited typical 40	Typical 2%	Si : 0.012 InP : 0.1 (At research stage only: Si+ organic material 0.0015 InP + Ti 0.01)
IOG. Concave Gratings (Siemens, IBM,..)	Laboratory solutions only up to 78	Down to 0.8	10 to 16	–10 to –30	Good	2% to 50%	0.1
FBG (QPS Tech., Bragg Photonics, Ionas, Alcatel, Lucent, HighWave..)	1 or a few channels available	0.4 (But lower spacing already tested in lab.)	1 per circulator	–25	Good	0%	0.01
3D Optics filters (JDS Fitel, E-Teck, Santec, OCA Corning Div, Dicon, HighWave...)	Available up to 16 (32)	Down to 0.8	1 to 6	–10 to –25	Good	0%	0.0005
3D optics grating (HighWave, JDS Fitel, Photonetics…)	Available up to 160 Feasibility more than *262 demonstrated*	Down to 0.2 (Feasibility down to 0.05 proved)	2 to 6	–30 to –55	High Typical 775	0%	Si :0.01 Si/Glass or aerial: 0.0003

References

[1] Smit, M. K., "New Focusing and Dispersive Planar Components Based on an Optical Phased Array," *Electron. Lett.*, Vol. 24, No. 7, 1988, pp. 385–386.

[2] Dragone, C., et al., "Efficient Multichannel Integrated Star Coupler on Silicon," *IEEE Photon. Technol. Lett.*, Vol. 1, 1989, pp. 241–243.

[3] Vellekoop, A. R., and M. K. Smit, "Four-Channel Integrated-Optic Wavelength Demultiplexer with Weak Polarization Dependence," *IEEE J. Lightwave Technol.*, Vol. 9, 1991, p. 310.

[4] Takahashi, H., et al., "Arrayed Waveguide Grating for Wavelength Division Multi/Demultiplexer with Nanometer Resolution," *Electron. Lett.*, Vol. 26, No. 2, January 18, 1990, pp. 87–88.

[5] Dragone, C., "Optimum Design of a Planar Array of Tapered Waveguides," *J. Opt. Soc. Amer. A.*, Vol. 7, No. 11, Nov. 1990, pp. 2081–2093.

[6] Dragone, C., et al., "Integrated Optics N × N Multiplexer on Silicon," *IEEE Photon. Technol. Lett.*, Vol. 3, Oct. 1991, pp. 896–899.

[7] Dragone, C., "An NxN Optical Multiplexer Using a Planar Arrangement of Two Star Couplers," *IEEE Photon. Technol. Lett*, Vol. 3, Sept. 1991, pp. 812–815.

[8] Takahashi, H., Y. Hibino, and Y. Nishi, "Polarization Insensitive Arrayed Waveguide Grating Wavelength Multiplexer on Silicon," *Opt. Lett.*, Vol. 17, No. 7, April 1, 1992, pp. 499–501.

[9] Inoue, Y., et al., "Polarization Mode Converter with Polyimide Half Wave Plate in Silica-Based Planar Lightwave Circuits," *IEEE Photon. Technol. Lett*, Vol. 6 (5), 1994, pp. 626–628.

[10] Verbeek, B. H., et al., "Large Bandwidth Polarization Independant and Compact 8 Channel Phasar Demultiplexer Filter," *OFC/IOOC'94 Technical Digest*, Post-deadline papers, San Jose, CA, Feb. 20–25, 1994, pp. 63–66.

[11] Vreeburg, C. G. M., et al., "An Improved Technology for Eliminating Polarization Dispersion in Integrated Phasar Demultiplexers," *IOOC/ ECOC 97* (*IEE Conf. Publ.* No. 448), Vol. 3, Edinburgh, Sept. 22–25, 1997, pp. 83–86.

[12] Inoue, Y., et al., "Athermal Silica-Based Arrayed-Waveguide Grating (AWG) Multiplexer," *ECOC' 97 Proc., Conference Publication IEE,* No. 448, 1997, pp. 33–35.

[13] Okamoto, K., "Fundamentals, Technology and Applications of AWG," *ECOC'98 Proc.*, Vol. 2, *Tutorial and Symposium papers*, Madrid, Sept. 1998, pp. 9–47.

[14] Takada, K., et al. "320-Channel Multiplexer Consisting of 100 GHz-Spaced Parent AWG and 10 GHz-Spaced Subsidiary AWGs," *Electron, Lett. (U.K.)*, Vol. 35, No. 13, May 10, 1999, pp. 824–826.

[15] Piri Mark. Dept. "Piri Expands AWG Product Line into the L-Band," *Piriodical*, No. 18, Feb. 2000.

[16] Yamada, H., et al., "Dispersion Resulting from Phase and Amplitude Errors in Arrayed Waveguide Grating Multiplexer/Demultiplexer," *Optics Lett.*, Vol. 25, Issue 8, April 15, 2000, pp. 569–571.

[17] Hill, K. O., et al. "Photosensitivity in Optical Fiber Waveguides: Application to Reflection Filter Fabrication," *Applied Physics Lett.*, Vol. 32, 1978, pp. 647–649.

[18] Kawasaki, B. S., et al. "Narrow-Band Bragg Reflectors in Optical Fibers," *Optics Lett.*, Vol. 3, 1978, pp. 66–68.

[19] Lam, D. K. W., and B. K. Garside, "Characterization of Single-Mode Optical Fiber Filters," *Applied Optics*, Vol. 20, 1981, pp. 440–445.

[20] Mayer, E., and D. Basting, "Excimer-Laser Advances Aid Production of Fiber Gratings, " *Laser Focus World*, April 2000, pp. 107–110.

[21] Bilodeau, F., "Photosensitization of Optical Fiber and Silica-on Silicon/Silica Waveguide," *Optics Lett.*, Vol. 18, 1993, pp. 953–955.

[22] Williams, D. L., et al., "Enhanced UV Photosensitivity in Boron Codoped Germanosilicate Fibers," *Electron. Lett.*, Vol. 29, 1993, pp. 45–47.

[23] Othonos, A., and K. Kalli, *Fiber Bragg Gratings*, Norwood, MA: Artech House, 1999.

[24] Russell St. J., P. L. Archambault, and L. Reekie, "Fiber Gratings," *Physics World*, October 1993, pp. 41–46.

[25] Malo, B., et al., "Apodized in-Fiber Bragg Grating Reflectors Photoimprinted Using a Phase Mask," *Electron. Lett.*, Vol. 31, 1995, pp. 223–225.

[26] Albert, J., et al., "Apodisation of the Spectral Response of Fiber Bragg Gratings Using a Phase Mask with Variable Diffraction Efficiency," *Electron. Lett.*, Vol. 31, 1995, pp. 222–223.

[27] Inoue, A., et al., "Optimisation of Fiber Bragg Grating for Dense WDM Transmission System," *IEICE Trans. Electron.*, Vol. E81 C, No. 8, August 1998, pp. 1209–1218.

[28] Delevaque, E., et al., "Optical Fiber Design for Strong Gratings Photoimprinting with Radiation Mode Suppression," *OFC'95 Proc.*, Post deadline paper #5, San Diego, CA, March 1995.

[29] Delevaque, E., and D. Pureur, "Composants Passifs à Fiber pour le Réseau WDM," *Journées d'Études SEE: Télécommunications Haut Débit et Multiplexage Optique WDM*, Marcoussis, France, March 23, 2000.

[30] Laude, J. P., et al., "Le Multiplexage de Longueurs d'Onde," *Opto 82 Conf. Proc.*, Paris: Masson Ed., 1982, pp. 144–147.

[31] Pelletier, E., and P. Bousquet, "Filtres Optiques Interférentiels Pour Multiplexeurs et Démultiplexeurs Destinés aux Télécommunications par Fibers," *ECOC' 82 Proc.*, Cannes, published by IEE London, Sept 1982, pp. 532–536.

[32] Minowa, J., and Y. Fujii, "Dielectric Multilayer Thin-Film Tilters for WDM Transmission Systems," *IEEE J. Lightwave Technol.*, Vol. LT-1, No.1, March 1983.

[33] Boitel, M., and A. Hamel, "Structures Multicouches Fabry-Perot Adaptées Aux Réseaux de Vidéocommunication," *Le Vide et Les Couches Minces*, No. 223, 1984, pp. 293–297.

[34] Stupik, P., "Bandpass Filter Design: Using Less Expensive Technology Without Sacrificing Performance," *The Photonics Design and Applications Handbook*, 1999, pp. H-285–290.

[35] Prieur, J., P. Davi, and V. Beaud, "DWDM Multidielectric Narrow-Band Optical Filters Manufacture," *SPIE Proc.*, Vol. 3408, No. 5, 1999, pp. 168–171.

[36] Furuta, H., K. Tsurata, and T. Miyamoto, "An Approach to Realize Sharp Cut-Off Characteristics Multi-ring Optical Filters," *Fukuoka University Review of Technological Sciences*, Vol. 62, No. 3, 1999, pp. 111–124.

[37] Furuta, H., K. Tsurata, and T. Miyamoto, "An Approach to Wide the Stop-Band of Multi Fabry Perot Optical Filters," *Fukuoka University Review of Technological Sciences*, Vol. 62, No. 3, 1999, pp. 107–110.

[38] Furuta, H., K. Nishikawa, and T. Miyamoto, "Wide Stop-Band Filters Consisted of Optical Directional Coupler Type Multi Resonators," *Fukuoka University Review of Technological Sciences*, Vol. 62, No. 3, 1999, pp. 125–134.

[39] Duck, G., "Performance and Application of Micro-optic Devices," *WFOPC '98 Proceedings (IEEE-LEOS)*, Pavia, Italy, Sept. 1998, pp. 50–56.

[40] Hutley, M. C., *Diffraction Gratings*, London: Acad. Press, 1982.

[41] Petit, R., *Electromagnetic Theory of Gratings*, Berlin: Springer-Verlag, 1980.

[42] Laude, J. P., and J. Flamand, "Herstellung von Bengungsgitten für Spektrometrie und Optoelektronik," *Feinwerktechnik und Messtechnick*, Vol. 94, No. 5, München, Germany: Carl Hanser Verlag, 1986.

[43] Lerner, J. M., et al. "Diffraction Gratings Ruled and Holographic: A Review," *SPIE*, Vol. 24014, 1980, pp. 82–88.

[44] Lerner, J. M., "Aberration Corrected Holographically Recorded Diffraction Gratings," *SPIE*, Vol. 24013, 1980, pp. 72–81.

[45] Laude, J. P., European patent 332790, filed March 18, 1988, et Nevière M., Maystre D. and J. P. Laude, "Perfect Blazing for Transmission Gratings," *J. Opt. Soc. Am. A*, Vol. 7, No. 9, Sept. 1990, pp. 1736–1739.

[46] Detaille, M., M. Duban, and J. P. Laude, "Réalisation de Réseaux de Diffraction Pour le Satellite Astronomique D2B," *Perspectives 91*, Jobin Yvon documentation, Feb. 1976.

[47] Lerner, J. M., and J. P. Laude, "New Vistas for Diffraction Gratings," *Electro-Optics*, May 1983, pp. 77–82.

[48] Maystre D., J. P. Laude, and P. Gacoin, "Gratings for Tuneable Lasers: Using Multidielectric Coatings to Improve Their Efficiency," *Appl. Opt.*, Vol. 19, No. 18, Sept. 15, 1980, pp. 3099–3102.

[49] Laude, J. P., et al., "Multiplexeur de Longueurs d'Onde à Micro-Optique Pour Fibers Unimodales," *Sixièmes Journées Nationales d'Optique Guidée*, Issy, France, March 20–21, 1985.

[50] Aoyame, K., and J. Minowa, "Low Loss Optical Demultiplexer for WDM Systems in the 0.8 μm Wavelength Region," *Appl. Opt.*, Vol. 18, No. 16/15, August 15, 1979, pp. 2854–2836.

[51] Watanabe, R., R. K. Nosu, and Y. Fujii, "Optical Grating Multiplexer in the 1.1–1.5 μm Wavelength Region," *Electron. Lett.*, Vol. 16, No. 3, 1980, p. 107.

[52] Tomlinson, W. J., "Wavelength Multiplexing in Multimode Optical Fibers," *Appl. Opt.*, Vol. 16, No. 8, August 1977, pp. 2180–2194.

[53] Metcalf, B. D., and J. F. Providakes, "High-Capacity Wavelength Demultiplexer with a Large-Diameter Grin Rod Lens," *Appl. Opt.*, Vol. 21, No. 5, March 1982, pp. 794–796.

[54] Seki, M., et al., "20-Channel Micro-optic Grating Demultiplexer for 1.1–1.6 μm Band Using a Small Focusing Parameter Graded-Index Rod Lens," *Electron. Lett.*, Vol. 18, No. 6, March 18, 1982, pp. 257–258.

[55] Tomlinson, W. J., "Wavelength Division Multiplexer," *U.S. patent*, Doc. 4, III, 524, Sept. 1978.

[56] Finke, G., A. Nicia, and D. Rittick, "Optische Kugellinsen Demultiplexer," *Optische Nachrichtentechnik*, Vol. 37, 1984, pp. 346–351.

[57] Laude, J. P., "High Brightness Monochromatic Using Optical Fibers," *Conference Opto Proc.*, Paris: ESI Ed., 1980, p. 60.

[58] Laude, J. P., and J. Flamand, "Un Multiplexeur-Démultiplexeur de Longueur d'Onde (Configuration Stimax)," *Revue Opto*, No. 3, Feb. 1981, pp. 33–34.

[59] Mannschke, L., "Microcomputer Aided Design and Realization of Low Insertion Loss Wavelength Multiplexer and Demultiplexer," *SPIE Proc.*, Vol. 399, 1983, pp. 92–97.

[60] Mannschke, L., "A Multiplexer/Coupler with Tapered Graded-Index Glass Fibers and a Grin Rod Lens," *ECOC'84 Proc.*, IEE Ed., 1984, pp. 164–165.

[61] Laude, J. P., et al., "STIMAX, a Grating Multiplexer for Monomode or Multimode Fibers," *ECOC'83, Proc.*, Geneva: Elsevier Science Pub., 1983, p. 417.

[62] Laude, J. P., *Wavelength Division Multiplexing*, London: Prentice Hall, 1993, p. 65.

[63] Watanabe, R., et al., "Optical Demultiplexer Using a Concave Grating in the 0.7–0.9 μm Wavelength Region," *Electron. Lett.*, Vol. 16, No. 3, 1980, p. 106.

[64] Fujii, Y., and J. Minowa, "Cylindrical Concave Grating Utilizing Thin Silicon Chip," *Electron. Lett.*, Vol. 17, No. 24, Nov. 26, 1981, pp. 934–936.

[65] Kita, T., and T. Harada, "Use of Aberration Corrected Concave Grating in Optical Demultiplexing," *Appl. Opt.*, Vol. 22, No. 6, March 15, 1983, pp. 819–825.

[66] Koonen, A. M., and A. Wismeijer, "Optical Devices for Wavelength Division Multiplexing Systems," *Philips Tel. Rev.*, Vol. 40, No. 2, July 1982, pp. 102–110.

[67] Watanabe, R., et al., "Optical Multi/Demultiplexer for Single-Mode Fibers Transmission," *IEEE J. Quantum Electronics*, Vol. QE 17, No. 6, June 1981, pp. 974–981.

[68] Leboutet, A. et al., "Wavelength Division Multiplexing in the 1.5 μm Window: An Installed Link," *Electron. Lett.*, Vol. 20, Sept. 27, 1984, pp. 834–835.

[69] Hegarty, J., et al., "Low-loss Single-Mode Wavelength Division Multiplexer with Etched Fiber Array," *Electron. Lett.*, Vol. 20, No. 17, August 16, 1984, pp. 685–686.

[70] Laude, J. P., "Les Multiplexeurs MonoModes à Réseaux de Diffraction: Largeurs Comparées Des Fonctions de Transmission Spectrales de Composants de 2 à 20 Voies," *Opto'90Proc, Paris, ESI Ed.*, 1990, p. 480.

[71] Laude, J. P., et al., "Results Obtained with 12 Multiplexers Using Single-Mode Fibers," *EFOC LAN 90 Proc.*, Boston: Information Gatekeeper, Inc. Ed., 1990, pp. 156–159.

[72] Wisely, D. R, "High Performance 32 Channels HDWDM Multiplexer with 1 nm Channel Spacing and 0.7 nm Bandwidth," *SPIE*, Vol. 1578, 1991, pp. 170–176.

[73] Laude, J. P., *Wavelength Division Multiplexing*, London: Prentice Hall, 1993.

[74] Laude, J. P., "New Miniaturized Diffraction Grating DWDM Components—Athermal High-Channel Count, 200 GHz to 20 GHz Spacing," IEEE Interactive Workshop on Fiber Optics, Optoelectronics and Photonics, Assembly, Packaging and Manufacturing, Open Forum Contribution, Reprint not available, Vail, CO, Sept. 15–17, 1999.

[75] Laude, J. P., "New Athermal Very Dense Wavelength Division Multiplexer," *Proc. ECOC 2000*, Vol. 3, Paper P2.10, Munich, Sept. 3–5, 2000.

[76] Green, P., *Fiber Optics Networks*, Englewood Cliffs, NJ: Prentice Hall, 1993, pp. 123–129.

[77] Takato, N., et al., "Silica-Based Integrated Optics Mach-Zehnder Multi/Demultiplexers Family with Channel Spacing of 0.01–250 nm," *IEEE J. of Selected Areas in Com.*, Vol. 8, No 6, Aug. 1990, pp. 1120–1127.

[78] Toba, H., et al., "100 Channel Optical FDM Transmission: Distribution at 622Mb/s over 50 km Using a Waveguide Frequency Selection Switch," *Electron. Lett.*, Vol. 26, No. 6, March 1990, pp. 376–377.

[79] Takato, N., et al., "128 Channel Polarization Insensitive Frequency-Selection-Switch Using High-Silica Waveguides," *IEEE Photon. Technol. Lett.*, Vol. 2, No. 6, 1990, pp. 441–443.

[80] Campbell, E., "DWDM a Feasible Alternative for Bandwidth-Hungry Broadband Networks," *Lightwave*, June 2000, pp. 102–110.

[81] LeBlanc, et al., "Passband Broadening of Integrated Arrayed Waveguide Filters Using a Multimode Interference Coupler," *Electron. Lett*, Vol. 32, No. 5, Feb. 1996, pp. 449–451.

[82] Boerk, H., et al., "Passband Flattening of Phasar WDM Using Input and Output Star Couplers Designed with Two Focal Points," *OFC'97 Technical Digest*, 1997, pp. 302–303.

[83] Rigny, A., et al., "Double-phased Array for a Flattened Spectral Response," *ECOC'97 Conf. Pub. IEE*, No. 448 , Vol. 3, 1997, pp. 79–86.

[84] Laude, J. P., Patent FR 97 05737, 1997.

[85] Dragone, C., "Efficient Techniques for Widening the Passband of a Wavelength Router," *IEEE. J. of Lightwave Technol.*, Vol. 16, No. 10, Oct. 1998, pp. 1895–1906.

[86] Laude, J. P., and S. Louis, "A New Method for Broadening and Flattening the Spectral Shape of the Transmission Channels of Grating Wavelength Division Multiplexers (WDM) and Routers," *OECC'98 Proc.*, Chiba, Japan, July 1998.

4

Sources and Wavelength Converters for DWDM

4.1 Introduction

The light sources used in DWDM are light-emitting diodes (LEDs), semiconductor lasers, or glass-doped lasers of different types. Some of these sources are wavelength tunable. The emitters can be directly or externally modulated. For high bit rates an external modulation is generally necessary. Bit rates in excess of 40 Gbps can be obtained from a CW source followed by an external modulator [1, 2]. Publications on external modulation are numerous. For instance, typical examples of high-speed LiNbO3 modulators and a typical application are reported in [1, 2], respectively. But we will not further discuss external modulation, which is outside of the scope of this book. WDM channels can be made from well-separated spectral slices selected in the spectrum of LEDs, superluminescent diodes, doped fiber amplifiers, or other coherent sources. All optical processing in the future telecommunication networks will need wavelength converters that can shift an optical signal from one wavelength to another. In this chapter, we review the main light sources and wavelength-converter techniques for DWDM.

4.2 Semiconductor Lasers

4.2.1 Laser Material

In direct-bandgap semiconductors, electron and hole populations with the same average momenta can be created by an electrical excitation. They can generate light in the recombination of the electrons and holes. "Compound semiconductors" with two or more elements in a crystalline structure equivalent to the structure of silicon can be used. The first problem is to get a semiconductor with large bandgap energy. A second problem is to match the lattice constant of the atoms to that of the substrate. In the III-V material Gallium Arsenide (GaAs) the bandgap is 1.42 Ev (Figure 4.1). This corresponds to a lasing wavelength of 905 nm. Other III-V semiconductor materials with Ga, In (III) and As, P (V) can be created showing other bandgaps corresponding to the wavelengths used in telecommunication. However, the lattice constant needs to be matched in any case to that of the substrate. In a quaternary compound such as InGaAsP, adding phosphorus atoms raises the bandgap but also reduces the lattice, as the atom P is smaller than the atom As that it partly replaces. The lattice can be matched by adding "large" In atoms replacing smaller Ga atoms.

InGaAsP/InP can provide varying bandgaps from 1.2 to 1.6 μm suitable for transmission on silica fibers. An InP substrate is generally used. However, a Si/InP substrate can be manufactured in specific cases. In this process the problems of lattice-constant mismatches between Si and InP is solved by specific methods, for instance a low temperature growth of a first InP layer on the Si substrate followed by the growth of a thick InP layer [3].

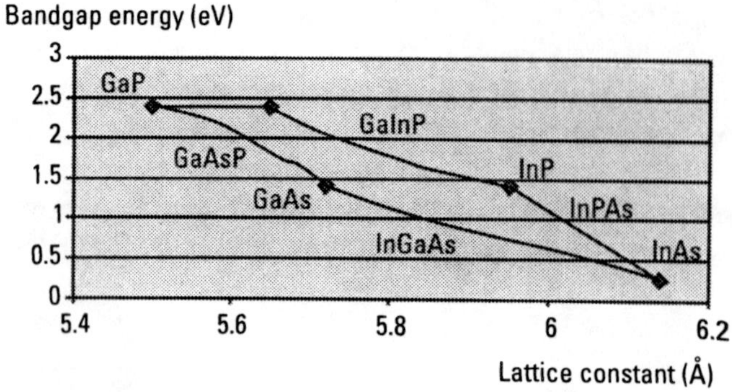

Figure 4.1 Bandgap energy versus lattice constant with III-V materials.

4.2.2 Quantum Well Lasers

The main problems with past devices were related to heating. In order to reduce the threshold current and then in order to reduce heating problems, the gain region was made thinner. However, below a junction thickness of about 0.2 μm in the common heterostructure the threshold current no longer decreases but increases again, due to diffraction losses in the narrow width of the gain layer. The quantum well (QW) design, first demonstrated in 1978 by Holonyak and now used in almost all devices, is a solution to this problem. In a QW structure the electron-hole pairs are confined in a thin junction region, and the photons emitted in the recombination are confined in a thicker region. The electron-hole confinement layer(s) can be made less than 50 nm thick, causing the bandgap to split into discrete subbands that help to further lower the threshold, and improve gain and coherence. The lattice-matching constraints are also relaxed with such thin layers.

In MQW lasers the threshold can be quite small (for example, a few mA in 670-nm GaIn P/AlGaInP ridge-waveguide lasers for short-range data transmission via optical fiber [4], down to 1 mA in narrow-beam laser arrays at 1.3 μm for parallel interconnections [5]).

4.2.3 Quantum Dot Lasers

Quantum dot (QD) lasers [6] have extremely small physical dimensions and low atom-like density of states associated to a strong confinement of electrons and holes. Compared to QW lasers, they can have low threshold-currents with weaker temperature sensitivity. The tuning range of QD lasers can be broad making them potentially useful for WDM applications. QD lasers with emission wavelengths of 1,300 nm and shorter wavelengths have been reported in different places [7]. QD lasers with a continuous tuning range from 1,400 to 1,650 nm became available recently. QD lasers are good candidates for ultra-short pulse generation (for example, 7ps, 7 GHz in [8]).

Potentially these lasers have an ultra-low frequency chirp so they will keep a sharp line-width at high bit rates.

In addition, an amplifier medium can be created by dispersing QDs in an optical waveguide: In January 2001, researchers at Università di Trento and at Università di Catania in Italy reported amplification with silicon QDs in a silica matrix on silicon (for more information, see pavesi@science.unitm.it).

4.2.4 Edge-Emitting Semiconductor Lasers

4.2.4.1 FP Lasers

The early laser diodes used FP cavities: the two mirrors of the cavity being obtained from cleaving the external facets of the chip (Figure 4.2). In the manufacturing process, the active strip (generally a MQW structure) needs to be etched. The etching (generally the reactive ion etching (RIE) process) is followed by a selective regrowth.

InP–based FP MQW lasers can obtained using an in-situ etching technique avoiding RIE that may damage the interface [9]. In this process, the active strip of a semi-insulating-buried heterostructure laser is made using 2-chloropropane as etchant in a MOCVD reactor, followed by a selective Fe-doped InP regrowth.

The FP devices oscillate in several longitudinal modes corresponding to the different FP peaks of reflection located in the broad gain spectrum of the active material. These modes are emitted in a bandwidth of about 2 nm at half maximum. Consequently these lasers are not suitable for DWDM. Different methods can be used to get a single-frequency (single-mode longitudinal) operation, including distributed-feedback (DFB), distributed-Bragg reflector (DBR), cleaved-coupled cavity (C^3), or external wavelength selective feedback (Figure 4.3).

4.2.4.2 DFB Grating Lasers

These structures were proposed in the early 1970s. The first DFB effect was demonstrated on dye lasers. But soon the principle was applied to semiconductor lasers. Today, DFB lasers are used in almost all installed DWDM networks. In these lasers a grating is formed along the active layer (Figure 4.4). The wavelength at which the feedback occurs is given by the grating step. A guiding layer transparent to the laser wavelength is grown over the grating.

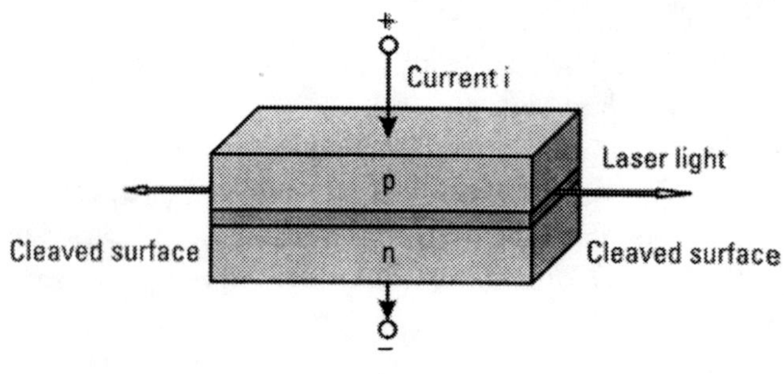

Figure 4.2 FP laser.

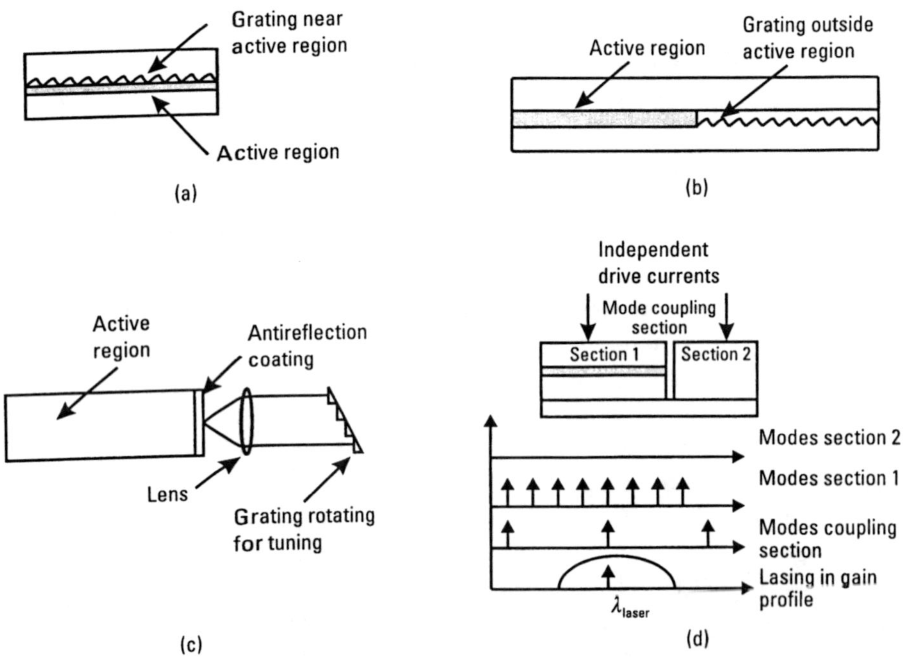

Figure 4.3 Single-frequency semiconductor lasers with four elementary designs: (a) DFB, a grating replaces the two mirrors of the cavity providing a feedback "distributed" along the active region; (b) DBR, the cavity "mirrors" consist of one grating on the semiconductor outside the active region and a cleaved facet; (c) external cavity laser; and (d) cleaved coupled-cavity laser.

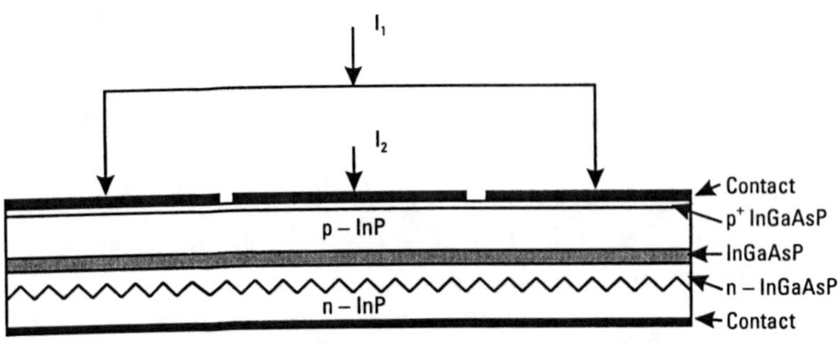

Figure 4.4 Multisection tunable DFB laser for 1.5 μm.

The grating coupling coefficient has a strong influence on the threshold current, on the quantum efficiency, on the high temperature performance, on the analog distortion, and on the SM lasing conditions. F. Pusa, et al. have proposed a method to extract this coefficient from measurements of the spontaneous emission spectrum of fabricated MQW-DFB lasers at 1.3 and 1.55 μm [10].

For each wavelength such that $\lambda = 2a\ n_{eff}/k$ (where a is the grating period, k is an integer, and n_{eff} is the effective index in the guiding structure) a feedback coupling through the grating is made possible. The lasing can occur if this wavelength is within the active material gain band. In general, for higher stability on single longitudinal-mode DFB lasers, the grating is made of two periodic structures of identical periods with a quarter-wavelength phase shift between the two structures. DFB lasers with this phase shift but also using asymmetrically pitch-modulated gratings (typical periods a = 240.5 nm and a – Δa = 240 nm) have been proposed in order to improve the slope-efficiency [11].

DFB lasers for DWDM with the frequencies specified on the ITU grid in the C band are commercially available from different manufacturers. A laser with a MQW electroabsorption (EA) modulator can be integrated on the same chip. Such light source modules demonstrated 40 Gbps NRZ operations [12]. They are suitable for 10-Gbps bit-rate transmissions over 40 to 100 km and for 2.5 Gbps over several hundreds of kilometers [13, 14].

A strong demand exists for L-band high-speed devices. Recently L-band EA-modulators integrated DFB-LD covering an 8-nm wavelength range on a single wafer were demonstrated. DWDM transmission at 10.66 Gbps per channel over 400 km was reported [15].

All DFBs developed today use variations of the QW active structure. In [16] the authors analyze the effects of nonhomogeneous carrier distribution in 1.55-μm lasers with seven 4.5 nm, 1% strained quantum-wells in which the top three wells are periodically etched. They show how to optimize such a device from a trade-off between high yield and low-threshold current density.

High-power (typically 100 mW) and high-fiber coupling efficiencies have been obtained with DFB-MQW single longitudinal-mode lasers at 1.55 μm. (See, for example, [17, 18]). A very-highpower and narrow line-width at 1.3 μm has important applications in analog cable-television broadcasting. Power above 220 mW maintaining single-longitudinal-mode operation has been obtained with DFB-MQW ridge waveguide structures [19].

For use in metropolitan area networks and gigabit Ethernet links such a high power is not necessary but there is an obvious demand for uncooled devices with direct modulation at high bit rates. InGaAlAs MQW lasers with a high SM yield at 1.3 μm can be designed. For example, in [20] a DFB laser was designed with 10 QWs and a reversed-mesa ridge waveguide structure. It can be modulated at 10 Gbps at 85°C and have a few milliwatts of power. Similar devices are reported in [21, 22] and in other papers.

Several DFB lasers can be integrated on a planar lighwave circuit (PLC) with a N to 1 star coupler to form a multiwavelength DWDM source. In [23] eight lasers at 200-GHz-spacing are integrated on a silica planar circuit. The authors demonstrate 2.5-Gbps operation with low thermal interference between channels.

Simultaneous fabrication of DFBs with different wavelengths on a single wafer is possible [24]. For instance, in [25] the manufacturing process of a set of 40 wavelength DFB-LDs and their characteristics are described. These lasers operate over a range of 75 nm from 1,518.2 to 1,593.2 nm. All threshold currents are below 12 mA. DFB arrays have been integrated with a power combiner and optical amplifiers [26]. Semiconductor amplifiers on InP can be efficiently integrated with an AWG [27]. The wavelengths of these devices can be locked to the same intracavity element. This guarantees a precise wavelength positioning [27, 28].

4.2.4.3 DBR Lasers

In the DBR, the grating is also integrated on the chip but the grating(s) is (are) placed outside of the active region. This simplifies the epitaxy. The device may use two gratings or only one grating and a reflection on a cleaved facet on the other side of the cavity. In a typical DBR tunable structure the index of the grating section can be controlled, and there is a phase section control between the grating section and the gain section (Figure 4.5). The wavelength can be tuned by the refraction index change by current injection

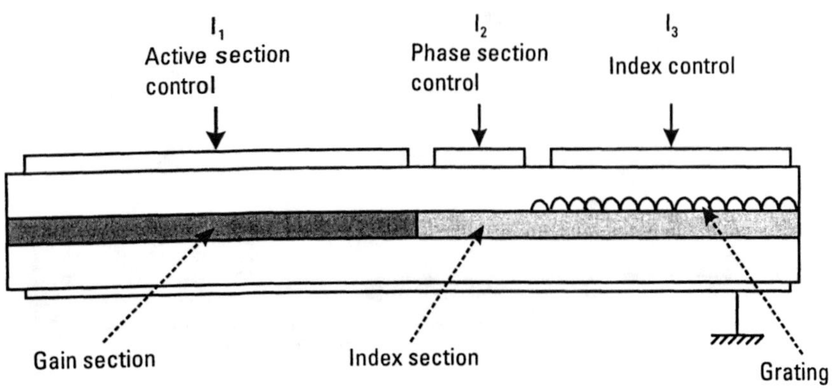

Figure 4.5 Typical DBR laser.

or thermal control. Several DBR lasers and electroabsorption modulators can be integrated for DWDM applications [29]. Various types of DBR lasers have been commercialized. They often use additional structures such as C^3 cavities (see Section 4.2.4.4) or super-structure gratings (see Section 4.2.4.5) that have a larger tunability than the conventional DBR lasers.

4.2.4.4 C^3 Lasers

C^3 lasers are designed with three sections with independent drive currents. Changing the gain controlled by the carrier density in the active section and independently changing the wavelength with an index variation controlled by the current in a different section are made possible. A third section is used for a matched optical coupling between these two sections. The laser emission can be obtained on a wavelength corresponding to modes of the three cavities being in phase somewhere in the gain profile. The C^3 lasers can be made tunable with a control of the optical lengths of the three cavities. This design can be combined with the DBR design.

4.2.4.5 Grating-Coupled Sampled Reflector Lasers

Of course many variants of coupled cavity devices exhibit two, three, or four integrated sections. The grating-coupled sampled reflector (GCSR) laser incorporates four sections. The tunability can be improved in these particular laser cavity structures. They can provide a large tuning range on any wavelength over the entire C band. The four sections "superstructure-grating" (SSG) or "sampled grating" (SG) DBR can cover a range of 30 to 100 nm [30–41], (Figure 4.6).

The sampled-grating distributed-Bragg-reflector laser consists of an active region, a phase region, and two SSG regions with one Bragg grating on each side of the device. These sections have comb-like reflection properties with different spacing between the peaks. By controlling the currents in these two sections, two peaks can be aligned and the laser oscillates on the

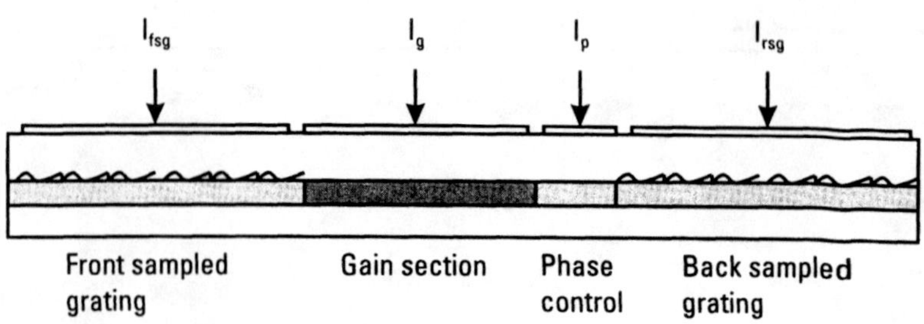

Figure 4.6 Four-section tunable SSG laser.

corresponding wavelength. The alignment can be obtained in a wide tuning range with only a small shift of the combs. In [38], F. Delorme gives the characteristics of the gratings of a widely tunable 1.55-μm SSG DBR: "The two reflectors consist of a periodic succession of grating bursts followed by uncorrugated regions. The length of the grating areas is 3.5 μm for both reflectors, and the uncorrugated complementary section length, different for both sides, is 28.5 and 30.5 μm, leading, respectively, to 32- and 34-μm SG period." The frequencies of the SSG DBR tunable lasers can be controlled by an external demultiplexer [40]. These lasers will become increasingly important in wavelength-agile DWDM optical networks.

4.2.4.6 External Cavity Lasers

The two reflectors of the cavity consist of a cleaved facet on one side and an external wavelength selective reflector (generally a grating) on the other side. These lasers, with a long cavity, have the narrowest line-width and can be easily tuned over the entire gain spectrum of the active material. They are widely used in instrumentation. However, as telecommunication sources, their reliability would be questionable until progress could be made on new microelectro-mechanical (MEM) designs now under development. Another potentially interesting design is the hybrid integration of multiwavelength lasers using gratings UV written with different periods on a planar waveguide [42] (Figure 4.7). The active chips are "flip-chip" bonded on terraces formed

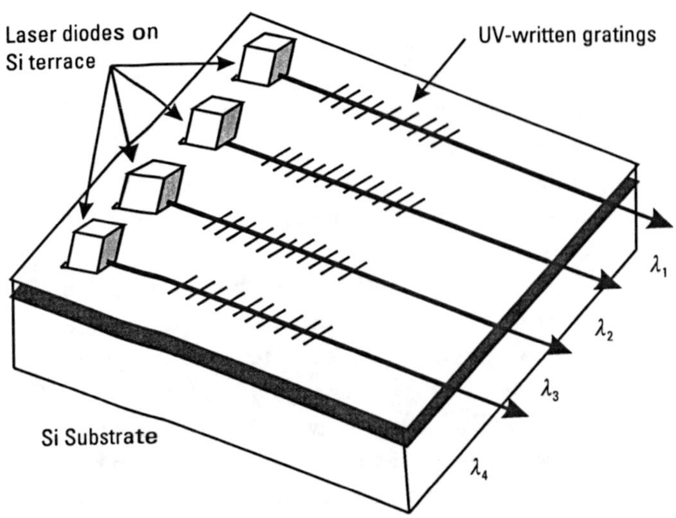

Figure 4.7 Hybrid integrated multiwavelength laser. (*After:* K. Okamoto [42].)

under an RIE process on a silicon substrate in front of optical waveguides with different periods Bragg gratings written under an UV beam. The cavity is shorter and is, in principle, more stable than equivalent devices with classical gratings.

Concurrently, a fiber grating can be used as the external wavelength selective feedback element [43, 44]. A hybrid semiconductor laser with an erbium fiber-based external cavity, on which a Bragg grating was written, has been proposed. This laser demonstrates a high SM stability [45].

4.2.5 Vertical-Cavity Surface Emitting Lasers

4.2.5.1 Principle and Main Practical Characteristics

Vertical-cavity surface emitting lasers (VCSELs) (Figure 4.8) are based on a QW active region inside a short cavity consisting of multiple-layer quarter-wavelength-stacked mirrors (DBRs). The typical threshold current of VCSELs is on the order of 1 mA. They generally emit a circular diffraction-limited light beam with a low numerical aperture. Single-transverse-mode operation can be obtained with different techniques.

The QW active layers are inside the optical cavity between DBRs. The higher order transverse-modes can be eliminated by an increase of losses using an annular etched-surface relief on the top facet [Figure 4.8(a)], a longer optical cavity [Figure 4.8(b)], or a tapered oxide aperture [Figure 4.8(c)]. The gain can also be preferentially provided to the fundamental mode with a specific central quantum gain region [Figure 4.8(d)]. (A reference on selective oxidation and additional references are available in [46, 47].)

VCSELs have shown important improvements over the last decade. They can have low submilliamp threshold currents (with proton or preferably oxygen implantation for current confinement [48, 49]), and a high electrical-to-optical conversion efficiency in excess of 15%. They emit in low divergence circular beam.

VCSELs at 1,550 nm are hampered by thermal problems on InP (i.e., the significant threshold current dependence on temperature), and the low thermal conductivity of the Bragg mirrors. However different solutions were proposed allowing to increase CW operation up to 90°C [50, 51].

Generally, the feedback sensitivity is not larger than in DFB laser diodes. These sources can be made tunable [52]. They can be integrated with refractive microlenses [53, 54]. Today, much work is devoted to microcavity effects and photonic bandgap structures including QD and quantum wire configurations.

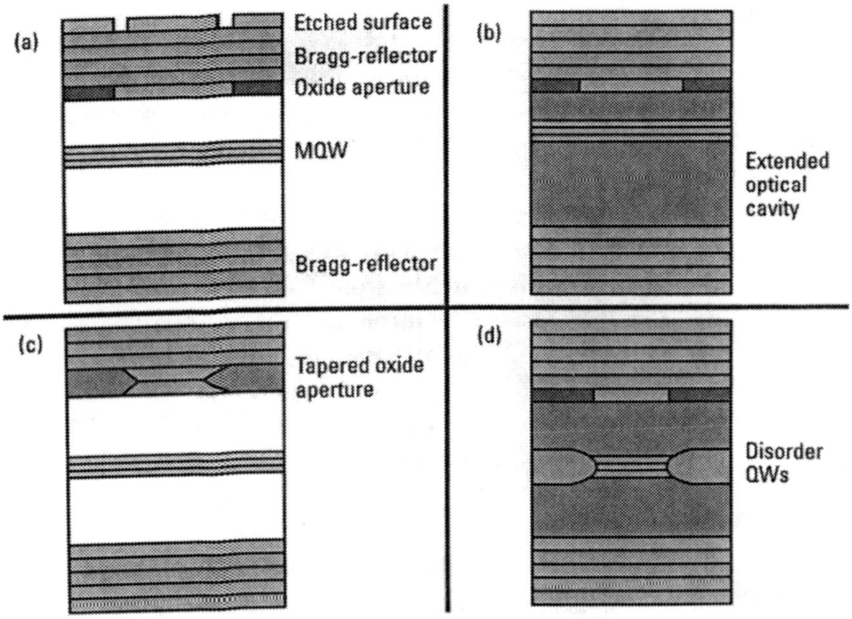

Figure 4.8 Different VCEL structures with: (a) annular etched-surface relief on the top facet, (b) extended optical cavity, (c) tapered oxide aperture, and (d) specific central quantum gain region.

VCSEL high-speed modules are commercially available from different manufacturers. One manufacturer claimed in 1998 [55] to be able to provide AlGaAs-based (near 850 nm) VCSELs for use in high-speed data-link modules with lifetimes exceeding 10^7 hours MTBF, and modulation speeds exceeding 1 Gbps. Of course, advanced VCSELs allowing WDM transmissions at 10 Gbps and above, and reducing the leakage current by various current aperture implantations, are studied in different laboratories (wavelength at 850 nm [56], at 1,550 nm [57], and at 825 nm [58]). For instance, a WDM transmission at aggregate capacity of 40 Gbps over four wavelengths at 814.7, 822.1, 828.0, and 835.0 nm over 310m of multimode fiber was recently demonstrated [58].

4.2.5.2 Temperature Effects on the Wavelength in VCSEL

There are two main effects: cavity length variation with index variation, and gain spectrum shift. In strain-compensated InGaAsP/InP multiquantum wells VCSEL for 1,550 nm, the corresponding wavelength drifts were shown to be $d\lambda/dT = 0.1$ nm/°C and 0.6 nm/°C, respectively [59].

4.2.5.3 VSCEL Arrays

Much work was devoted to VSCEL arrays. They are one of the key components in optical interconnect devices including crossbar switches. Much work was devoted to this subject in Europe [60] and elsewhere [61–64] during the last 10 years.

The VCSEL is also a natural candidate for coherent coupling of two or more elementary sources in laser arrays. This is a way to increase the power in a low divergence beam array. Researchers from the University of Wisconsin presented at the IEEE International Semiconductor Laser Conference (Monterey, California, Sept. 24–28, 2000) such devices with 4×4 arrays emitting coherently 2 mW of power at 980 nm, a wavelength that is suitable for Er fiber amplifier pumping. According to the authors, 40×40 arrays as well as emissions at 1.3 or 1.5 μm would be feasible.

4.2.5.4 Microcavity Structures

Semiconductor microcavities including special VCSELs, photonic-band crystals, and whispering-gallery-mode cavities are good candidates for very-low-threshold very-small source integration [65].

Among them, particular microstructured emitters using photonic crystals do not allow spontaneous emission to be emitted in all directions but to be channeled in adequate modes. New microlasers and microLEDs based on photonic crystals with very-high quantum efficiencies can be obtained [66]. For instance, an efficient SM laser operation at 10 Gbps was obtained into a standard FP diode laser with two-dimensional photonic crystals implemented on either side of the laser waveguide [67].

4.2.6 Wavelength Tunability in Semiconductor Lasers

Following is a comparison of the tunability of the different technologies discussed in the previous sections.

4.2.6.1 Continuous Wavelength Tuning

The external cavity wavelength selective feedback gives a wide continuous tunability of the semiconductor in the gain bandwidth. This solution is often used in instrumentation applications, however its thermal and mechanical stability may be questionable in network applications, with perhaps the noticeable exception of VSCELs that can be tuned externally with MEMS in hybrid microconfigurations.

At the present time, DFB and DBR lasers are generally preferred as tunable sources in most network applications. However, their continuous wavelength tuning range is limited to values significantly lower than the gain bandwidth [68].

DFB and DBR laser wavelengths can be controlled by the temperature and by the currents in different sections, with tuning of the index of refraction resulting in the tuning of the wavelength: $\Delta\lambda/\lambda = \Delta n/n$. The maximum values of $\Delta n/n$ and the switching time delay are given in Table 4.1 [38].

In theory, this gives a tuning range extending from 10 to 17 nm. In practice, lasers with 20 channels and more in the C band were announced to become commercially available in 2000 [69–71]. Every day the laboratories of different manufacturers demonstrate reliable module-packaged devices with more and more channels. (For instance, three-sections gain-coupled DFB lasers integrated with a semiconductor optical amplifier (SOA) with 33 channels over 12.8 nm on the 50-GHz ITU grid [72].)

Systematic investigations of aging of two-section butt-jointed DBR lasers, grown in three metal-organic vapor phase epitaxy (MOVPE) steps, tunable over 12 nm, have been performed [73]. It was concluded that the reliability was high enough for practical implementation in telecommunication systems.

4.2.6.2 Discontinuous Wavelength Tuning

We have seen that special configurations using interference effects in different sections of the cavity, SSG, or sampled-grating distributed-Bragg reflectors (SSG-DBR and SG-DBR) allow tuning on extended ranges in the discontinuous mode. SSG-DBR and SG-DBR seem to be the most attractive devices. Multiple-section lasers with sampled grating DBR designs allow a wide tuning range up to 75 nm and more.

Table 4.1
Relative Refraction Index Modulation and Switching Speed

	Index Modulation $\Delta n/n$	Switching Speed
Carrier injection	~0.01	Few ns
Thermal heating	~0.01	Few ms
Electrorefraction	~0.002	~100 ps

(*Source:* [38].)

4.3 Glass-Doped-Based Lasers with Narrow Line-Widths

4.3.1 Principle

So far, semiconductor lasers have been used as CW or pulsed-light sources in multiwavelength networks. Er fiber lasers with different wavelength selective feedback devices such as diffraction gratings, AWGs, Bragg gratings, and FP cavities have been used for lasing operations in the 1.5-μm wavelength region. Generally, due to a long cavity, these lasers have many closely spaced longitudinal modes in a broad spectral range in which the wavelengths could be selected. It is difficult to solve the problem of residual noise due to the selection in each channel of several modes.

However, fiber-ring lasers with a highly reflective FBGs permit stable narrow-line-width operations at relatively high power. For example, a ring laser with a 3.7-m long Er^{3+}-doped fiber demonstrated a line-width <0.056nm (limited by resolution of test instrumentation) at 19 mW of output power with 70 mW of pump power [74].

A special resonator with two Bragg gratings, one being $\pi/2$-phase-shifted designed for longitudinal-mode discrimination, has been proposed to achieve single-frequency operation, with an output power of several milliwatts, on a laser having a relatively long cavity (30 cm in the experiment) [75].

A single-frequency fiber laser can be obtained more easily using a shorter cavity. The use of ytterbium-codoped erbium (Yb/Er) phosphate-glass or alumino-phosphosilicate fibers or planar Yb/Er waveguides [76], which are capable of high pump absorption over short distances with small absorption of the Er ions at the laser wavelength, allows a reduction of the cavity length. The pump absorption in doped fibers can be about 0.2 dB/mm at 980 nm with a typical stimulated emission absorption of 30 dB/m. The pump absorption is about 100 times larger than what can be obtained with Er fibers. For example, in 1997 efficient and tunable single-frequency operations over 11 and 25 nm in both 1,535 and 1,550 nm were demonstrated with novel short cavity Yb/Er fiber grating lasers with fiber gain section of 3 to 25 mm [77]. Another example of an Er/Yb codoped fiber single frequency laser with an output power of −10 dBm was successfully demonstrated in 1998. The cavity was a fiber loop of about 11 cm and a fiber grating was used for wavelength selection [78].

As an example of a compact waveguide laser, a research team obtained 2 mW of output power at 1,535 nm, with a signal-to-noise ratio of 53 dB, from a subcentimeter-long waveguide laser. This laser used a planar 9-mm-long Er/Yb waveguide obtained by ion exchange in phosphate glass, and a 10-mm-long FBG. The same team reported a pump efficiency of 27% for another 43-mm-long waveguide laser [79] (Figure 4.9).

The wavelength can be made tunable, the grating period being modified by compression or stretching of the fiber, or multiple-fiber lasers emitting at different wavelengths being used. (An example of multiple-fiber laser is given in [80].)

4.3.2 Fiber-Based Short-Pulse Laser Sources

A way to design a pulsed multiwavelength light source is to use the multiwavelength spectrum of mode-locked semiconductors or ring-fiber lasers. Such periodic lasers can be used as short-pulse sources in soliton transmission.

Active mode-locking based on a nonlinear-optical loop mirror (NOLM) gives a low interpulse-stream timing jitter (900 fs) [81].

Mode locking with a fiber acoustooptic frequency shifter has been used in a soliton laser [82]. This laser was configured in a ring geometry (cavity length 25m) using 7m of Er/Yb fiber. It delivered 18 ps transform-limited laser pulses at an output power of 20 mW.

In principle, fiber-based devices can offer high power ultrashort (≤ 5ps) pulse generation with broadband tunability. The active mode locking seems by far the best technique for telecommunication applications. However, the practical field usage of mode-locked fiber lasers may be delayed by their relative complexity [83].

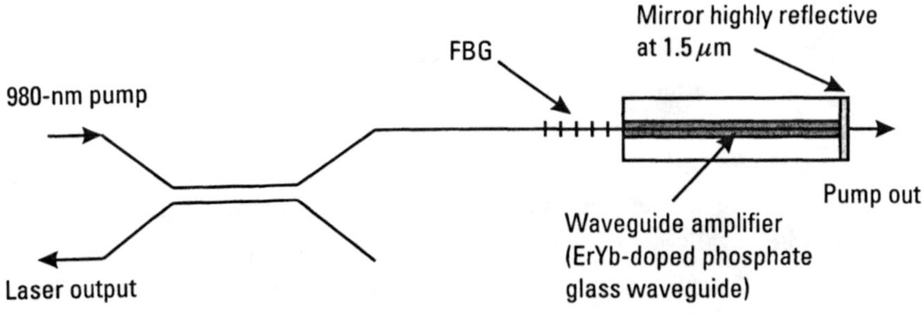

Figure 4.9 Typical short-glass waveguide laser. (*After:* D. Barbier, et al. [79].)

4.4 Spectral Slicing of Sources

4.4.1 Principle

The use of WDM generally provides flexibility in the elaboration of solutions and the evolution of networks. But the cost saving related to reducing the number of fibers (this may result in further civil engineering, cable, and connection cost savings) is sometimes counterbalanced by extra costs resulting from more stringent specifications of optical sources. If the bit rate of each elementary signal is not too large and if the transmission lengths are short enough, it is nevertheless possible to use low-cost LEDs (up to some tens of megabits per second over a few kilometers; typical applications: process control networks, telephone links, etc.). The choice of LEDs with well-separated wavelengths is limited, but the LED spectral-slicing technique, that we will discuss now, allows a large increase in the number of channels. In this technique, a set of LEDs, identical or with adequate spectral shifts are typically connected to the multiplexer entrances. The multiplexer cuts well-separated spectral slices, one different slice for each entrance. For example, in the case of a grating multiplexer, the spectrum of each identical source connected to a given input fiber is imaged on the output fiber with a position shift, depending on the input fiber position in the object plane. A slice of wavelengths that is different for each input fiber is selected in the output fiber. Coherent superluminescent diodes [84], amplified spontaneous emission from doped fibers [85], supercontinuum (SC) generation in fibers [86], and femtosecond lasers [87] offer more coupled power with a high signal-to-noise ratio.

4.4.2 Typical Examples

The first reports on LED slicing in optical fiber networks (1982 to 1985) did not receive much attention. It was pointed out that for short-distance links and low bit rates, one could tolerate a specified spectral overlap or even multiplex signals from identical diodes. The multiplexing of 20 identical LEDs could be obtained with –45 dBm signals at the receivers for 2 km and 50/125 μm fiber links with the low power LEDs available at that time [88, 89]. This was not enough for telecommunication applications with the optical sources available at that time, but it was adequate for multiplexed optical sensor networks. Such a network for process control using only one LED and wavelength-sliced signals reflected from different optical sensors was described [90]. Framatome installed such systems in different industrial sites from 1985 to 1990. The specifications of a 42-channel multiplexer obtained with 1 nm slice-widths, cut in a LED emitting around 820 nm and used in a process control network with 100/140 μm fibers, were given in 1984 [91]. It soon became possible to use the LED-slicing technique on SM fibers in some telecommunication network applications: M. H. Reeve, et al. [92] described a four-channel network with 20-nm spacing between channels around 1,300 nm and at 2 Mbps. The use of this technique in the U.K. loop network has been reported by Hunwicks, et al. [93]. The use of superluminescent diodes allowed the extension of the spectral slicing up to 16 channels as early as 1990 [94].

The effect of source temperature variation (an example is given in Figure 4.10) was analyzed in detail by Hunwicks [95] at 1,300 nm.

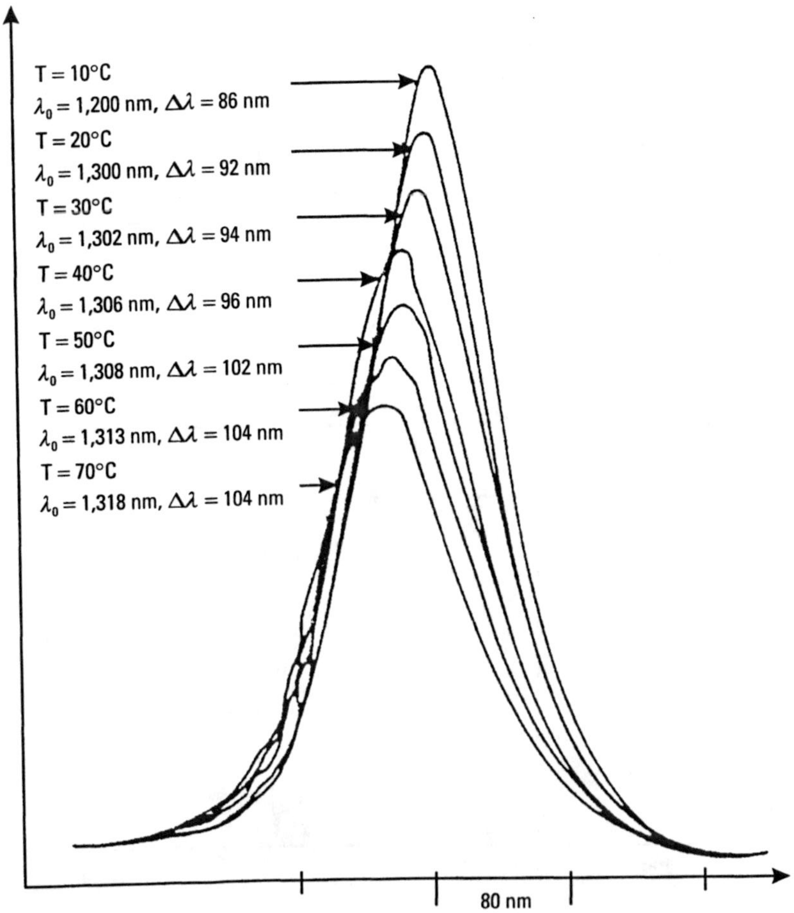

Figure 4.10 Typical emission spectrum at different temperatures of a 1,300-nm LED. (*Source:* CIT Alcatel.)

Considering a 0.54 nm/°C thermal shift from 0°C to 70°C, he showed that additional losses appear up to 3.8 dB for a four-channel multiplexed system, and 4.7 dB for 10 channels (SM fiber case). Over the last decade, much work has been devoted to the impact of the thermal drift on the optical crosstalk of spectrally sliced WDM networks with the conclusion that this technique can work without thermal stabilization under certain conditions in local access networks [96]. Calculations on interchannel interference effects and their influence on system specifications were published [97].

Bersiner and Rund [98] demonstrated bidirectional transmission in five channels over 8 km of standard GI fibers at bit rates of 167 Mbps per channel in the 1.3 μm domain using InGaAs avalanche photodiodes.

The characteristics of a typical multiplexer for LED slicing are given in [99] (Figure 4.11). For example, with 10 SM fiber channels, a spectral spacing between channels of 9 nm at 1,300 nm and a FWHM of the wavelength transmission functions of 1.6 nm on the multiplexer and 6 nm on the demultiplexer, the combined losses for the multiplexer and the demultiplexer are under 5 dB, and the optical crosstalk is lower than –31 dB.

The use of spectral slicing for wavelength-routed subscriber loops, with superluminescent diodes at 150 Mbps over 10 channels or 50 Mbps over 16 channels and 7 km in a first demonstration, and with LEDs over 4 channels, without optical amplification, was experimented by T. E. Chapuran,

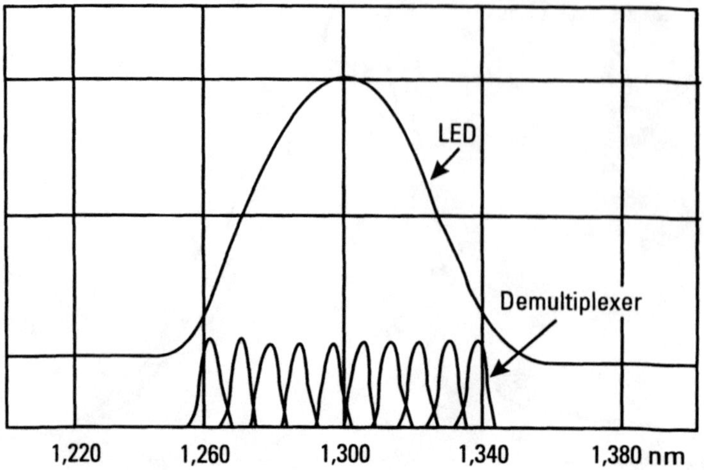

Figure 4.11 Characteristics of the emitter and passbands in the slices (fiber with mode diameter 10.8 and 47.5 μm between fibers).

et al. [84]. In a third experiment, these authors showed how erbium-doped fiber amplifiers could greatly alleviate the power-budget constraints of spectral slicing, as it is possible to keep the same performance in networks connecting splitter-based fiber-loop architectures to a broadband switching hub. These experiments show how the amplifiers can mitigate the effects of splitting losses as well as other sources of attenuation. G. J. Lampard [100] arrived at an equivalent conclusion: In networks using spectral slicing and optical amplification, the capacity can be increased to 30 channels operating at 60 Mbps.

A spectrum-sliced fiber amplifier source was used to transmit 15 downstream channels operating at 500 Mps each with LED's at the subscriber for upstream channels [101]. After about 20 km of SM fiber, the power is still acceptable (–43 dBm level).

4.4.3 Calculation of Spectral Filtering Losses

Let us assume that the transmission fiber is SM along with all entrance fibers of the multiplexer. We have seen that the multiplexer spectral transmission function is Gaussian if the system has no aberrations. We assume, in addition, that the diode spectrum can be approximated by a Gaussian function (Figure 4.12). These functions have the following expressions:

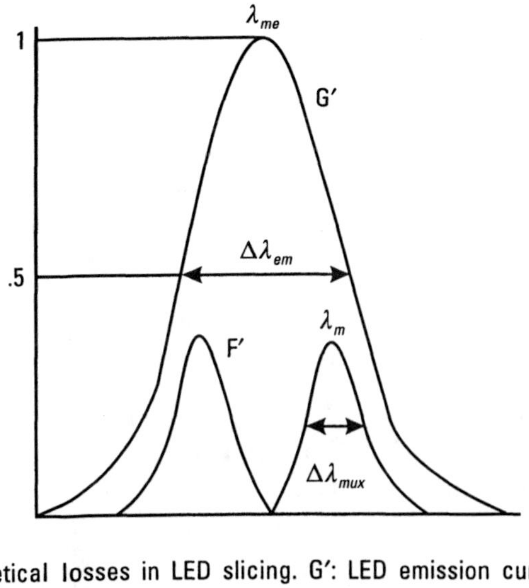

Figure 4.12 Theoretical losses in LED slicing. G′: LED emission curve, F′: Multiplexer channel transmission.

- For the multiplexer:

$$F' = \exp\left[-4\ln 2\left(\frac{\lambda - \lambda_{m}}{\Delta\lambda_{mux}}\right)^{2}\right]$$

- For the emitter:

$$G' = \exp\left[-4\ln 2\left(\frac{\lambda - \lambda_{e}}{\Delta\lambda_{em}}\right)^{2}\right]$$

In these equations, we introduced the FWHM, $\Delta\lambda_{mux}$ for the multiplexer and $\Delta\lambda_{em}$ for the emitter. λ_m is the wavelength at a specific passband maximum of the multiplexer and λ_e is the wavelength at the emitter maximum.

The transmission is given by the convolution product:

$$[F' * G']$$

If the demultiplexer is identical to the multiplexer, we will obtain a demultiplexer additional loss related to the cross correlation of the Gaussian functions of the multiplexer and of the demultiplexer. The total transmission is reduced by a factor:

$$\left[\frac{1}{\sqrt{2}}\right] = 0.7071$$

If the demultiplexer uses multimode exit fibers, the transmission functions are larger and the theoretical loss contribution of the demultiplexer is negligible.

4.4.4 Calculation of the Spectral Filtering Losses of Real Systems

Let us consider the actual case of Figure 4.13. Four channels are sliced in two LEDs (two channels in each LED) through an all-single-mode fiber multiplexer with the following characteristic ratio:

$$\frac{\Delta\lambda_{mux}}{\text{Channel spectral distance}} = 0.44$$

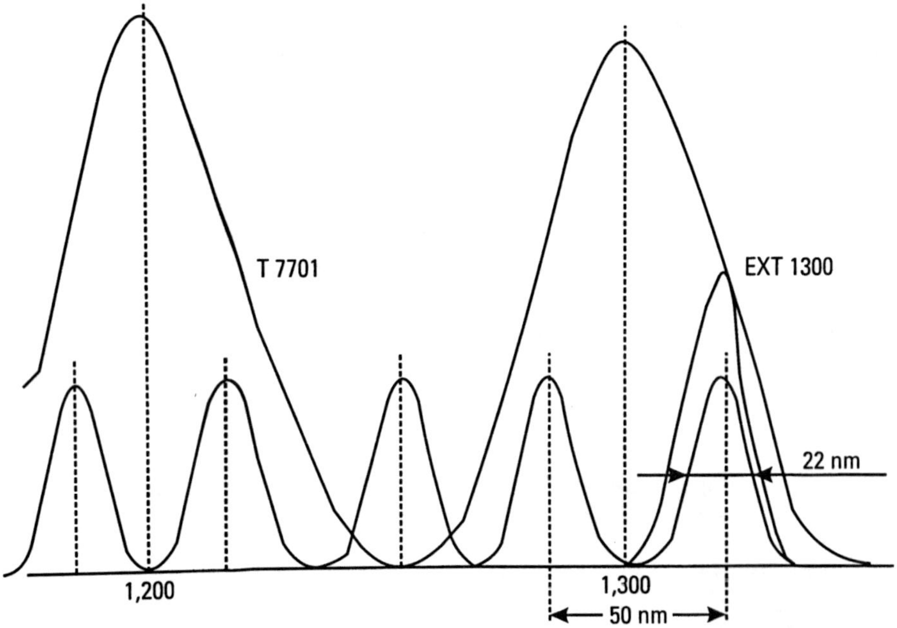

Figure 4.13 Typical spectral slicing multiplexing: Two LEDs, four channels. Upper curves: Emission spectra. Lower curves: Multiplexer transmission channels. On right, between upper and lower curves: Product of emission × transmission.

The emitter emission curve is approximated to a Gaussian function with a FWHM of $\Delta\lambda_{em} = 61$ nm; the multiplexer has a Gaussian transmission function with $\Delta\lambda_{mux} = 22$ nm. With $\lambda_m - \lambda_e = 25$ nm, the power in a slice is typically 10 dB below the launch power (including the component-defect losses).

Let us consider now M. H. Reeve, et al.'s device [92]. This is an SM, four-channel multiplexer at 1,270, 1,290, 1,310, and 1,330 nm, $\Delta\lambda_{mux} = 3.65$ nm, $\Delta\lambda_{em} = 63$ nm. For the worst channel, $\lambda_e - \lambda_m = 30$ nm, the multiplexing slicing losses are 16 dB. With a launch power of −26.7 dBm, a multiplexer and demultiplexer total excess loss of 9 dB, a fiber and splice loss of 2.4 dB, and a receiver sensitivity of −56 dBm, a margin of 2.4 dB is still available.

Let us now assume that the devices are those of [99]. The multiplexer is an all-single-mode fiber, 10-channel component. On the demultiplexer, multimode 50/125 μm fibers are used, so the "shaping" losses due to the demultiplexer can be neglected. The multiplexer and the demultiplexer have FWHMs of 1.6 and 6 nm, respectively. If an emitter with $\Delta\lambda_{em} = 90$ nm is used, a transmission T = 0.02 to 0.01 from the channel near the emitter maximum to the edge channels is obtained; this corresponds to 20 to17-dB losses. As powers of 10 to 20 μW (–20 dBm to –17 dBm) can be injected with typical diodes, –40 to –34 dBm can be obtained in the worst channel with such devices. With a receiver of –56.5 dBm (10^{-9} BER) sensitivity at 2 Mbps, a 16.5 to 22.5 dB margin remains available for the transmission line, including 4 to 6 dB of multiplexer plus demultiplexer typical loss.

In 1992, an LED specially designed for spectral slicing became available (mean wavelength: 830 nm, FWHM: 120 nm, power coupled in 400 μm core: 400 μW). Today, efficient superluminescent diodes at 1.5 μm, and other wavelengths, designed for spectrum slicing, are also available from different laboratories and manufacturers (1.5 μm, 200 mW in [102]).

4.4.5 Combining Spectral Slicing and TDM

This solution has been proposed for simple and efficient use in access networks (typically 10- to 20-km distances). A semiconductor amplifier-modulator monolithically integrated with a broadband LED is used to generate and amplify pulses at high bit rates. The short broadband pulses are amplified in an erbium-doped fiber amplifier (EDFA) and bit interleaved in a WDM/TDM modulation format through a static wavelength router equipped with passive delay lines ("loop-back multiplexing"). At the receiver end the signals are wavelength demultiplexed and detected with commercial receivers. Twenty-four channels with 200-GHz spacing were encoded at 50 Mbps per channel in an amplifier modulator after the router. An excellent bit-error-rate level in experimental access systems was demonstrated [103].

4.4.6 SC Lightwave Optical Sources: Coherent Sources for Spectrum Slicing WDM

The SC source was invented in 1970 [104], and has been used first in ultra-fast spectroscopy.

This is a coherent source; as such, it greatly reduces the spontaneous-emission beat noise that limits the spectral efficiency in the classical LED slicing [105]. In SC generation a high-power pump-pulse spectrum is broadened over a continuous frequency range while maintaining its pulse nature. A mode-locked laser generates short periodic pulses corresponding to a spectrum of equidistant frequencies. The temporal width of these pulses is reduced, and consequently the spectrum is broadened, by nonlinear effects in a dispersion-shifted fiber [106].

The effect can be analyzed using the Fourier transform (FT) relation between time and frequency. Of course, the FT is the "mathematical embodiment" of the uncertainty principle: localizing in time corresponds to spreading in frequency. A signal $f(t)$ corresponds to its transform F (ω):

$$\mathrm{F}(\omega) = \frac{1}{\sqrt{2\pi}} \int_{-\infty}^{-\infty} f(t)\left(\exp - i\omega t\right) \mathrm{d}t$$

A short pulse gives a broad frequency spectrum. For instance, a Gaussian pulse of temporal width $\Delta\tau$ at 1/e corresponds to a Gaussian optical-frequency spectrum of width $\Delta\nu$ at 1/e with the relation $\Delta\nu\Delta\tau = 1/\pi$ [107].

The FT of a sampled time function of period (t in the time domain is a periodic function of period 1/Δt in the frequency domain (and conversely). Short pulses of width $\Delta\tau$ emitted periodically at Δt time intervals correspond to periodic frequencies at $1/\Delta t$ frequency intervals in an envelope of width $\kappa/\Delta\tau$ in the frequency domain. The factor κ depends of the pulse shape.

In practice, the spectrum of mode-locked laser seed pulses has few discrete channels when it enters in the device and many over a wide frequency range at the output of the nonlinear fiber that reduces the pulse width and consequently broadens the spectrum. The distance between the different channels is that of the repetition rate.

In ultra-fast spectroscopy applications, mode-locked YAG SHG or color-center lasers were generally used. In telecommunications applications the pump can be a mode-locked fiber laser as in [108], in which pulses with a spectral width of about 70 nm are extracted from a 36.7-MHz femtosecond laser and chirped through a dispersive fiber.

Recently, picosecond pulses from amplified laser diodes at around 10-GHz repetition rates were used with good stability and reliability. After spectral expansion in a nonlinear fiber, the different channels are separated into multiple pulses of different carrier wavelengths. For instance, an SC source with over 1,000 channels, at 12.5-GHz spacing from 1,500 to 1,600 nm, ranged from −15 to −23 dBm, has been obtained from a diode laser mode-locked at 12.5 GHz [109]. The fiber was a polarization maintaining dispersion-flattened fiber, designed such that the wavelength-dispersion characteristics are in the form of a convex function, with a decrease from the input to the output of the fiber.

4.4.7 Comparison of Different Technologies

We reviewed the main different technologies used for wavelength sources. A continuous tunability over a wide spectral range has substantial benefits in DWDM applications. The investment in inventory can be reduced, as few tunable lasers can replace multitudes of fixed-wavelength lasers. Arrays of fixed-wavelength lasers instead of individual lasers can also be used. Above all, a wide tunability allows routing and add/dropping on a large scale, while a fast tuning will also allow full packet switching. Table 4.2 compares the main typical characteristics of different solutions that are well established or still at the research stage. It may help in choosing the most appropriate solution for a given application. However, it was not possible to be fully exhaustive as the figures change every day.

4.5 Wavelength Converters

4.5.1 Introduction

Different techniques have been proposed to perform wavelength conversion. An ideal converter should have the following performances:

- High signal-to-noise ratio;
- Low-chirp of output signal;
- High data rate;
- Fast setup time;
- Bit-rate transparency to intensity modulated signals, and more generally amplitude frequency and phase transparency to signals with any modulation format;
- Cascadability;
- Signal reshaping;
- Polarization insensivity;
- Large wavelength range with conversion to both shorter and longer wavelengths;
- Simultaneous conversion of a set of input wavelengths;
- Ability to give identical input and output wavelength when no conversion is required;
- Easy implementation and low power consumption.

Table 4.2
Comparison of Various Different Technologies for DWDM Tunable Sources

	Typical Output Power MW	Side Mode Suppression Ratio (SMRS) dB	Typical Wavelength Tunability	Remarks
DFB	2/30 (lower with tunable DFB)	33/38	5 nm	Well-established manufacturing process. 40-Gbps integrated electroabsorption modulator reported.
DFB laser arrays with multiplexer	2/10	40/50	Discrete fixed channels	8 channels 3.2-nm spacing available. More than 40 channels at research stage.
DBR	2/10	30/35	17 nm	Well established manufacturing process
C^3	2/10	20/25	16/30 nm	
GCSR	10/15	30/40	40/60 nm 100 nm feasible	Fast 5/30-ns response time Relatively complex drive Not fully continuous tuning
Microexternal cavity laser	10/20	50	40 nm	Low switching speed Vibration sensitivity
VCSEL with MEMS	1	35/40	14 nm and more	Fast 3/10 Gbps 850/1,300 nm 1,500 nm is becoming available
Spectrum sliced	–5/–10	30	Discrete channels	40 channels 0.4-nm spacing at research stage

Different solutions are available. None of them has all these desired features. We will concentrate on:

- Optoelectronic conversion;
- All-optical conversion;
- Cross-modulation techniques including cross-gain modulation (XGM), and XPM;
- Coherent techniques including FWM and difference frequency generation (DFG).

The reader is referred to [110–120] and to the other references given in the following text.

4.5.2 Optoelectronic Conversion

The optical signal is first converted to an electronic signal by a photoreceiver. After amplification, regeneration (2R regeneration: amplifying and reshaping; or 3R regeneration: amplifying, reshaping, and retiming), and buffering in the electronic domain, the signal drives the input of a tunable laser or the input of an integrated laser array delivering the required wavelength from a set of predetermined wavelengths (Figure 4.14).

The emitter may be a single laser with a single wavelength, a tunable laser, multiwavelength lasers integrated monolithically with a space switch [121], or a set of discrete DFB lasers with a coupling switch [122].

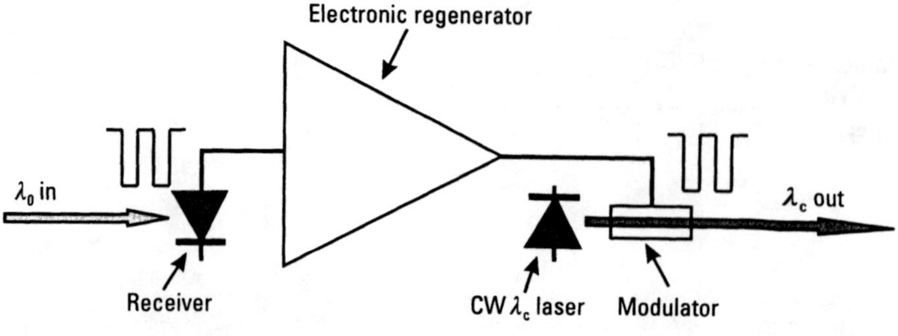

Figure 4.14 Optoelectronic converter. The wavelength λ_0 of the incoming signal is converted to the wavelength λ_c of a CW laser. The signal is applied to an optical modulator after electronic regeneration. The modulation is transferred to λ_c. In other versions, this signal directly modulates the laser.

4.5.3 XGM

This wavelength conversion principle is based on nonlinearities in SOAs. In an SOA the amplification is correlated with carrier depletion in the amplifier material. The modulated input signal saturates the gain of the SOA that in turn modulates the transmission of a CW source set at the assigned wavelength (Figure 4.15). The incoming signal is translated to the CW source wavelength and inverted (a maximum of the incoming signal corresponds to a minimum of the converted signal output). A higher extinction ratio conversion for very low input power can be obtained, replacing the SOA by an integrated SOA/DFB laser [123].

XGM is very simple and can operate at high bit rate (the carrier dynamic allows conversion bit rates of 80 Gbps and more). At very-high bit rates, it is necessary to reduce the signal-pattern effects induced in the SOA by the finite gain recovery time. This can be done using a filtering by a fiber grating on the converted signal in XGM [124] or XPM [125].

With an interferometric compensation, even an ultra-fast XGM conversion of picosecond RZ pulses [126], or an ultra-high bit rate (168 Gbps) optical TDM XPM wavelength conversion [127], were made possible.

The XGM process is polarization independent. The main problems of practical devices remain the spectral chirp in the converted signal due to index modulation, and some degradation of the extinction ratio.

4.5.4 XPM

In these devices the index is modulated through the carrier density variation induced by the incoming signal in one of the arms of an interferometer. A phase difference between the two arms of the interferometer is created by the input signal at λ_i. A CW light at λ_c transmitted through the interferometer is

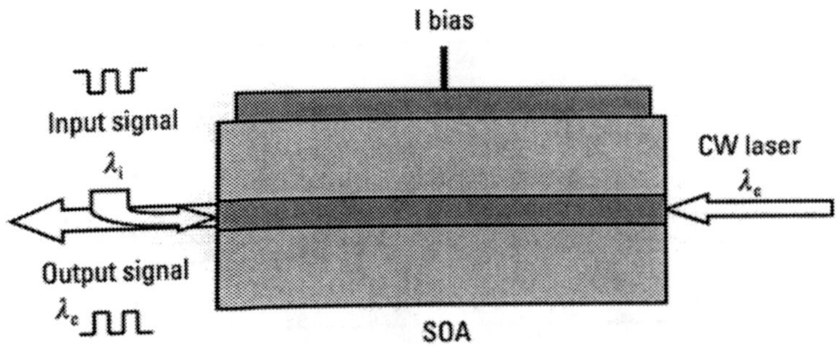

Figure 4.15 XGM in an SOA.

in turn modulated by this phase modulation. The interferometers are in general of MZ (Figure 4.16) [128–131], or of Michelson types (Figure 4.17) [132, 133]. L. H. Spiekman, et al. designed an MZ converter with a DFB source and a SOA preamplifier integrated on-chip (Figure 4.18).

All these interferometers can be made polarization insensitive. The chirp is generally very small because the phase modulation is obtained with very-small gain variations. The devices are fast, allowing bit rates well in excess of 10 Gbps. Copropagating schemes have the faster responses (see for instance [134]).

New configurations of XPM conversion using Kerr-induced phase modulation in optical fiber interferometers are under investigation [135]. A tunable range of 30 nm was demonstrated.

Conversions using NOLM configurations with SOAs or fibers have been reported. NOLMs based on silica fibers can potentially attain Tbps bit rates in NZ or NRZ format due to their ultra-fast optical nonlinearity [136, 137].

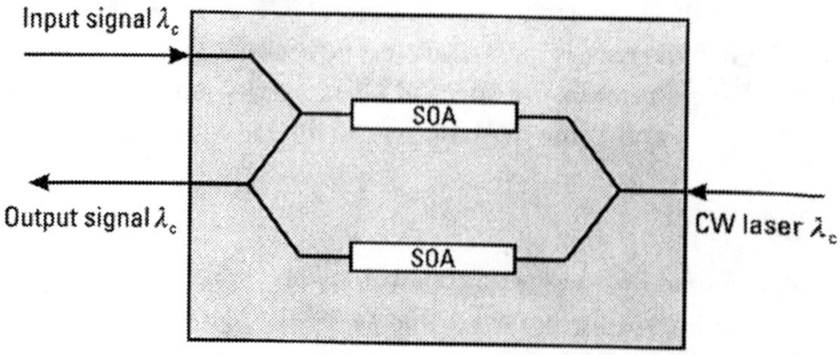

Figure 4.16 Wavelength conversion using a counter-propagating MZI configuration.

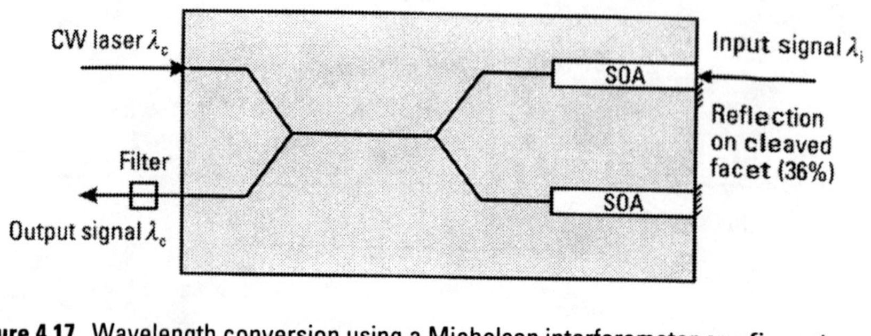

Figure 4.17 Wavelength conversion using a Michelson interferometer configuration.

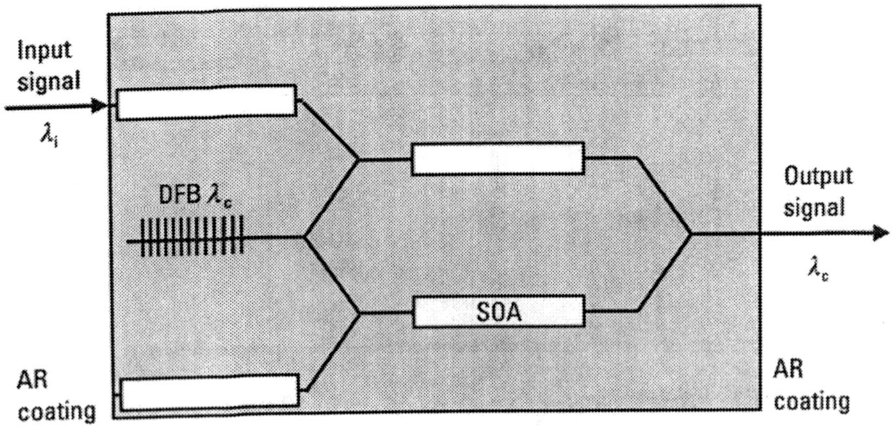

Figure 4.18 Wavelength converter with a DFB and SOA preamplifier monolithically integrated on-chip. (*After:* Spiekmann, et al. [138].)

XPM converters can be used for suppression of the accumulated noise of concatenated EDFAs [139]. One possible drawback is the limited input power dynamic. These converters are not transparent to the signal phase. But for digital signals cross-phase modulators are know to be among the best solutions for all-optical conversion.

4.5.5 FWM

In Section 7.4, we discuss FWM frequency generation in fibers as a possible source of crosstalk. However, this effect in nonlinear fibers or in SOAs can also be used for wavelength conversion [140, 141]. In FWM conversion, the input signal at frequency ω_i is added to a CW pump at frequency ω_p in a nonlinear optical medium. A signal at frequency $\omega_c = 2\,\omega_p - \omega_i$ is generated (Figure 4.19). The conversion efficiency can be high in a SOA with very long cavity [142].

FWM is a good candidate for fully transparent wavelength conversion, as it preserves both phase and amplitude. It allows simultaneous conversion of a set of multiple wavelengths to another set (see, for instance, [143]). A wide pump tunability is possible in SOA [144].

Conversion can be obtained over a large detuning band in special highly nonlinear fibers (34 nm in [145, 146]). Moreover as it gives a spectral inversion FWM can also be applied for dispersion compensation when used

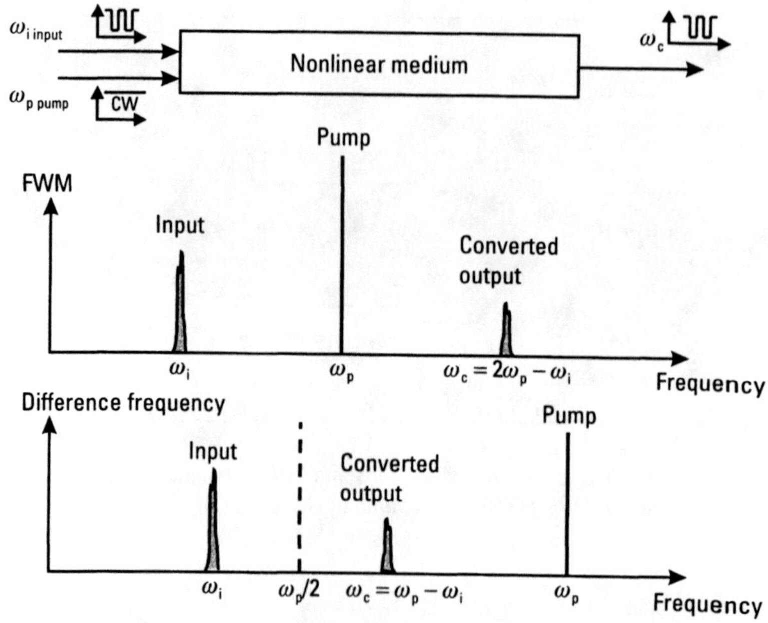

Figure 4.19 FWM and difference frequency generation for wavelength conversion.

at midspan. In order to get polarization insensitivity, two pumps or other compensations are required. An example is given in [147]. In conventional FWM converters the conversion range is limited to 15 nm at 10 Gbps [148]. More valuable information can be found in [149–154].

4.5.6 Difference Frequency Generation

DFG, also called three-wave mixing, allows for multichannel wavelength conversion with low crosstalk [155]. The converters are insensitive to polarization variations and are fully transparent. In [156] simultaneous DFG conversion in an AlGaAs waveguide of eight input wavelengths (1,546 to 1,560 nm, 2-nm spacing) to a set of eight output wavelengths (1,538 to 1,524 nm, 2-nm spacing) is demonstrated. The pump power was approximately 65 mW at 780 nm. In [157] a periodically poled LiNbO3 substrate is used. A multiring configuration is proposed to get a conversion without polarization sensitivity. A normalized conversion efficiency of 44%/Wcm^2 is obtained.

4.5.7 State of the Art in Wavelength Conversion in 2001

The optoelectronic method is a mature technology generally used for amplitude transparent conversion at bit rates up to 2.5 Gbps. For higher speed operations all-optical technologies are advantageous. At 10 Gbps and above the XPM may be preferred. XPM on SOA allows 2R regeneration of the signal. The coherent methods, still under development at the time of this writing, are very attractive. They allow ultra-fast conversions; they are potentially fully transparent; and they are able to convert simultaneously multiple wavelengths. For multichannel wavelength conversion DFG at the present time seems to be potentially the technique of choice. However, its efficiency and the pump tunability need to be improved.

References

[1] Hallemeir, P., et al., "Next Generation 10 Gb/s Lithium Niobate Modulator Components for RZ Based Transmission Techniques," *NFOEC'99, Tech. Proc*, Vol. 2, Chicago, Sept. 26–27, 1999, pp. 175–178.

[2] Brandon, E., et al., "1.28 Tbit/s (32 × 40 Gbit/s) Unrepeated Transmission over 250 km," *ECOC'2000 Proc.*, Vol. 4, Munich, Sept. 3–5, 2000, pp. 21–22.

[3] Yamada, M., et al., "1.55 μm Laser on Si Substrate," *NTT Opto-Electronics Laboratories Annual Report*, 1997, p. 30.

[4] Kuhn, J., et al., "Low-Threshold High Quantum Efficiency 670nm GaInP/AlGaInP Ridge-Waveguide Lasers," *LEOS'97 Proc., IEEE Lasers and Electro-Optics Society Annual Meeting*, Vol. 1, San Francisco, CA, Nov. 1997, pp. 20–21.

[5] Nakahara, K., et al., "A Low Threshold and Narrow-Beam Divergence N-Type Modulation-Doped Strained MQW Laser Array for Parallel Optical Interconnection," *ECOC'98 Proc.*, Vol. 1, Madrid, Sept. 20–24, 1998, pp. 71–72.

[6] Arakawa, Y., and K. Okamoto, "Foreword: Special Issue on Advanced Optical Devices for Next Generation High-Speed Communication Systems and Photonics Networks," *IEICE Trans. Electron*, Vol. E83, No. 5, p. 787.

[7] Shoji, H., et al., "Self-Formed In(Ga)As Quantum Dot Lasers," *LEOS'97, IEEE Proc.*, Vol. 2, San Francisco, CA, 1997, pp. 494–495.

[8] Xiaodong, H., et al., "Passive Mode-Locking in 1.3 μm Two Section InAs Quantum Dot Lasers," *Appl. Phys. Lett.*, Vol. 78, No. 19, 2001, pp. 2825–2827.

[9] Bertone, D., et al., "High Reliability, High Yield, High Modulation Bandwidth, Low Threshold Current 1.55μm MQW Laser by New In-Situ Etching Technique," *ECOC'98 Proc.*, Vol. 1, Madrid, Sept. 20–24, 1998, pp. 75–76.

[10] Pusa, F., et al., "An Automatic Method to Extract the Grating Coupling Coefficient in DFB Lasers: Evaluation of Different Types of Fabricated Lasers," *IOOC-ECOC'97 Proc.*, Vol. 2, No 448, IEE Ed., Edinburgh, Sept. 22–25, 1997, pp. 200–201.

[11] Sato, K., et al., "High-Slope-Efficiency, High-Single-Longitudinal-Mode-Stability DFB Laser Diodes with Asymmetrical-Pitch-Modulated (APM) Gratings," *ECOC'2000 Proc.*, Vol. 1, Munich, Sept. 3–5, 2000, pp. 133–134.

[12] Takeuchi, H., "Very High-Speed DFB-Modulators Integrated Light Sources," *OECC'97 Technical Digest*, Seoul, South Korea, July 1997, pp. 18–19.

[13] Takeuchi, H., et al., "10 Gb/s Light Source Module Containing MQW-Electroabsorption Modulator and DFB Laser," *NTT Opto-Electronics Laboratories Annual Report*, 1997, p. 13.

[14] Bregi, P., "Multicolor Lasers and Modules Open the Way to the Future," *Journée d'Étude SEE-Club 31 Télécommunications Optiques Haut Débit à Multiplexage Optique WDM*, Alcatel Marcoussis, France, March 23, 2000.

[15] Furushima, Y., et al., "In-Wafer Wavelength Distribution Controlled L-Band DFB/MODs for 10 Gb/s/ch Long-Haul DWDM Systems," *ECOC'2000 Proc.*, Vol. 1, Munich, Sept. 3–5, 2000, pp. 129–130.

[16] Champagne, A., et al., "The Effects of Non-Homogeneous Carrier Distribution on MQW DFB Laser Performance," *LEOS'97, Proc. IEEE Lasers and Electro-Optics Society Annual Meeting*, Vol. 1, San Francisco, CA, Nov. 1997, pp. 75–76.

[17] Jeon, H., et al., "High Power 1.55μm DFB Laser Diodes Integrated with Vertical Mode Expanders," *LEOS'97, Proc. IEEE Lasers and Electro-Optics Society Annual Meeting*, Vol. 1, San Francisco, CA, Nov. 1997, pp. 67–68.

[18] Chen, T. R., et al., "High Power 1550 nm DFB Lasers for Long Distance Video Transmission," *LEOS'97, IEEE Lasers and Electro-Optics Society Annual Meeting*, Vol. 1, San Francisco, CA, Nov. 1997, pp. 69–70.

[19] Ziari, M., and A. Mathur, "High Power and Narrow Line-Width 1.3 μm Distributed Feedback Lasers," *LEOS'97, IEEE Lasers and Electro-Optics Society Annual Meeting*, Vol. 1, San Francisco, CA, Nov. 1997, pp. 71–72.

[20] Aoki, M., et al., "85°C -10 Gb/s Operation of 1.3 μm In GaAlAs MQW-DFB Laser," *ECOC'2000 Proc.*, Vol. 1, Munich, Sept. 3–5, 2000, pp. 123–124.

[21] Sakaino, G., et al., "Uncooled and Directly Modulated 1.3 μm DFB Laser Diode for Serial 10 Gb/s Ethernet," *ECOC'2000 Proc.*, Vol. 1, Munich, Sept. 3–5, 2000, pp. 125–126.

[22] Takiguchi, T., et al., "1.3μm AlGaIn As Lasers with 12.0 GHz Relaxation Oscillation Frequency at 85°C for Gigabit Ethernet System," *ECOC'2000 Proc.*, Vol. 1, Munich, Sept. 3–5, 2000, pp. 127–128.

[23] Yanagisawa, M., et al., "A 2.5 Gb/s Hybrid Integrated MultiWavelength Light Source Composed of Eight DFB-LD's and an MMI Coupler on a Silica PLC Platform" *ECOC'98 Proc.*, Vol. 1, Madrid, Spain, Sept. 20–24, 1998, pp. 77–78.

[24] Kobayashi, K., and I. Mito, "MultiWavelength Lasers on a Chip," *LEOS'97, IEEE Lasers and Electro-Optics Society Annual Meeting*, Vol. 1, San Francisco, CA, Nov. 1997, pp. 65–66.

[25] Kudo, K., et al., "Over 75-nm-Wide Wavelength Range Detuning-Adjusted DFB-LDs of Different Wavelengths Fabricated on a Wafer," *OFC'97*, Post-deadline papers PD 15-2, Dallas, TX, Feb. 16–21, 1997.

[26] Zah, C. E., et al., "Wavelength Accuracy and Output Power of MultiWavelength DFB Lasers with Integrated Star Couplers and Optical Amplifiers," *IEEE Photon. Technol. Lett*, Vol. 8, 1996, pp. 864–866.

[27] Zirngibl, M., "Integrated WDM Light Sources," *IOOC-ECOC'97 Proc.*, Vol. 1, No. 448, IEE Ed., Edinburgh, Sept. 22–25, 1997.

[28] Soole, J. B. D., et al., "Multi-Strip Array Grating Integrated Cavity (MAGIC) Laser: A New Semiconductor Laser for WDM Applications," *Electron. Lett.* Vol. 28, 1992, pp. 1805–1807.

[29] Young, M. G., et al., "A 16 × 1 WDM Transmitter with Integrated DBR Lasers and Electroabsorption Modulators," *IEEE Photon. Technol. Lett.*, Vol. 5, 1993, pp. 908–910.

[30] Plastow, R., "Fast and Fine Wavelength Tuning of a GCSR Laser Using a Digitally Controlled Driver," *OFC 2000 Technical Digest*, Paper WM43, 2000.

[31] Jayaraman, V., et al., "Theory, Design, and Performance of Extended Tuning Range Semiconductor Lasers with Sampled Grating," *IEEE J. Quantum Electronics*, Vol. 29, No. 6, 1993, pp. 1824–1834.

[32] Ishii, H., et al., "Broad-Range Wavelength Coverage (64.4nm) with Superstructure-Grating DBR Laser," *Electron. Lett.*, Vol. 32, No. 5, 1996, pp. 454–455.

[33] Ougier, C., et al., "Sampled-Grating DFB Laser with 80 Addressable Wavelengths over 33nm for 2.5 Gb/s WDM Applications," *Electron. Lett.*, Vol. 32, No. 17, 1996, pp. 1592–1593.

[34] Rigole, P. J., et al., "Quasicontinuous Tuning Range from 1560 to 1520 nm in a GCSR Laser, with High Power and Low Tuning Currents," *Electron. Lett.*, Vol. 32, No. 25, 1996, pp. 2325–2354.

[35] Tohmori, Y., et al., "Over 100nm Wavelength Tuning in Superstructure Grating (SSG) DBR Lasers," *Electron. Lett.*, Vol. 29, No. 4, 1993, pp. 352–354.

[36] Lee, S., et al., "Integration of Semiconductor Laser Amplifiers with Sampled Grating Tunable Lasers for WDM Applications," *IEEE J. Quantum Electronics*, Vol. 3, No. 2, 1997, pp. 615–627.

[37] Coldren, L. A., et al., "Properties of Widely-Tunable Integrated WDM Sources and Receivers," *LEOS'97, IEEE Lasers and Electro-Optics Society Annual Meeting*, Vol. 1, San Francisco, CA, Nov. 1997, pp. 331–332.

[38] Delorme, F., "Widely Tunable 1.55-μm Lasers for Wavelength-Division-Multiplexed Optical Fiber Communications," *IEEE J. Quantum Electronics*, Vol. 34, No. 9, Sept. 1998, pp. 1706–1716.

[39] Mason B., "Widely Tunable Semiconductor Lasers," *ECOC'2000 Proc.*, Vol. 2, Munich, Sept. 3–5, 2000, pp. 157–158.

[40] Ishii, H., F. Kano, and Y. Yoshikuni, "Wavelength Stabilization of a Superstructure-Grating DBR Laser for WDM Networks," *IOOC-ECOC'97 Proc.*, Vol. 1, No. 448, IEE Ed., Edinburgh, Sept. 22–25, 1997, pp. 103–106.

[41] Kano, F., et al., "Semiconductor Light Sources for WDM Network Systems," *NTT Review*, Vol. 10, No. 1, Jan. 1998, pp. 21–29.

[42] Okamoto, K., "Fundamentals, Technology and Applications of AWGs," *ECOC' 98 Tutorial*, Vol. 2, Madrid, 1998, pp. 44–45.

[43] Paoletti, R., et al., "10Gb/s Ultra-Low Chirp 1.55μm Directly Modulated Hybrid Fibre-Grating-Semiconductor Laser Source," *IOOC-ECOC'97 Proc.*, Vol. 1, Edinburgh, Sept. 22–25, 1997, pp. 107–110.

[44] Takagi, T., et al., "Fiber-Grating External–Cavity Laser Diode Module for 2.5 Gb/s Dense WDM Transmission," *ECOC'98 Proc.*, Vol. 1, Madrid, Sept. 20—24, 1998, pp. 81–82.

[45] Loh, W. H., et al., "Novel Hybrid Single Frequency Semiconductor Laser with Erbium Fibre-Based External Cavity," *ECOC'95 Proc.*, Vol. 1, Brussels, Sept. 17, 1995, pp. 135–138.

[46] Choquette, K. D., et al., "Selective Oxidation for Record CW 640–660 nm AlGaInP VCSELs," *CLEO'95 Technical Digest*, Post-deadline papers CPD5-1, Baltimore, MD, 1995.

[47] Choquette, K. D., "Selectively Oxidized VCSELs Go Single-Mode," *Laser Focus World*, May 2000, pp. 251–254.

[48] Michalziz, R., K. J. Ebeling, "Modeling and Design of Proton Implanted UltraLow Threshold Vertical Cavity Laser Diodes," *IEEE J. Quantum Electronics*, Vol. 29, 1993, pp. 1963–1974.

[49] Quian, Y. et al., "Low Threshold Room Temperature CW 1.3 μm Single-Bonded Vertical-Cavity Surface-Emitting Lasers Using Oxygen-Implanted Confinement," *OFC'97* Post-deadline papers PD 14-2, Dallas, Texas, Feb. 16–21, 1997.

[50] Amann, M. -C., et al., "High-Performance Vertical-Cavity Surface-Emitting Lasers for Telecommunication Wavelengths," *ECOC'2000*, Post-deadline papers 3.3, Munich, Sept. 3–5, 2000.

[51] Hall, E., et al., "88°C, Continuous-Wave Operation of 1.55μm Vertical-Cavity Surface-Emitting Lasers," *ECOC'2000*, Post-deadline papers 3.4, Munich, Sept. 3–5, 2000.

[52] Wipiejewski, T., et al., "Tunable Extremely Low Threshold Vertical-Cavity Laser Diodes," *Photon. Technol. Lett.*, Vol. 5, 1993, pp. 889–892.

[53] Strzelecka, E. M., et al., "Integration of Vertical-Cavity Laser Diodes with Refractive Microlenses," *CLEO'95 Technical Digest*, Vol. 15, Baltimore, MD, 1995, pp. 194–195.

[54] Du, T., et al., "Directional Beam Control Using On/Off-axis High-Efficiency Diffractive Optics Integrated on Substrate-Emitting Vertical Cavity Lasers," *CLEO'95 Technical Digest*, Vol. 15, Baltimore, MD, 1995, pp. 195–196.

[55] Morgan, R., et al. "Reliable and Low Cost Commercial VCSEL Modules for Optical Data Links," *OECC'98 Tech. Digest*, Chiba, Japan, July 1998, pp. 30–31.

[56] Lear, K. L., et al., "Small and Large Modulation of 850nm Oxide-Confined Vertical Cavity Surface Emitting Lasers," *OSA TOPS*, Vol. 15, *Advances in VCSELs*, C. Chang-Hasnain (ed.), 1997.

[57] Stevens, R., et al., "High Speed Modulation Characteristics of Long Wavelength Vertical Cavity Lasers Based on an Integrated InP Bragg Reflector," *ECOC'98 Proc.*, Vol. 1, Madrid, Sept. 10–24, 1998, pp. 79–80.

[58] Michalzik, R., et al. "40 Gb/s Coarse WDM Data Transmission with 825 nm Wavelength VCSELs over 310 m of High-Performance Multimode Fiber," *ECOC'2000 Proc.*, Vol. 4, Munich, Sept. 3–5, 2000, pp. 33–34.

[59] Piprek, J., "Analysis of Temperature Effects on 1.55 μm Vertical-Cavity Lasers," *LEOS'97 Proc.*, Vol. 2, San Francisco, CA, Nov. 10–13, 1997, pp. 428–429.

[60] Neefs, H., "Achievements 1996–2000: Optoelectronics Interconnects for Integrated Circuits," *MEL-ARI OPTO*, European Commission DG Information Society Ed, June 2000, pp. 32–33.

[61] Kurokawa, T., et al., "VCSEL Based Smart Pixels Toward Parallel Optoelectronic Processing Systems," *LEOS'97 Proc.*, Vol. 2, San Franscisco, CA, Nov. 10–13, 1997, pp. 11–12.

[62] Ozguz, V., "Silicon Complexity Required for VCSEL Based Smart Pixels," *LEOS'97 Proc.*, Vol. 2, San Francisco, CA, Nov. 10–13, 1997, pp. 13–14.

[63] Gulden, K., et al., "Individually Addressed VCSEL Arrays," *LEOS'97 Proc.*, Vol. 2, San Francisco, CA, Nov. 10–13, 1997, pp. 15–16.

[64] Tateno, K., et al., "0.85 μm Vertical Cavity Surface Emitting Laser Array on a GaAs (311)B Substrate," *NTT Opto-Electronics Laboratories Annual Report*, 1997, p. 29.

[65] Baba, T., "Photonic Band and Whispering Gallery Mode Emitters," *LEOS'97 Proc.*, Vol. 2, San Francisco, CA, Nov. 10–13, 1997, pp. 3–4.

[66] Krauss, T. F., and R. M. De La Rue, "The Potential of Photonic Microstructures in Optoelectronics," *IOOC-ECOC'97, Tutorial Notes*, IEE Ed., Edinburgh, Sept. 22–25, 1997.

[67] Massara, A. B., et al., "10Gb/s Single-Mode Operation of 2D Lattice Distributed Reflector Laser," *ECOC'2000 Proc.*, Vol. 1, Munich, Sept. 3–5, 2000, pp. 131–132.

[68] Amann, M. C., "Wavelength Tunable Laser Diodes and Their Applications," *Trends in Optical Fibre Metrology and Standards, NATO ASI Series E.*, Vol. 285, Dordrecht, The Netherlands: Kluwer Academic Publishers, 1995, pp. 217–240.

[69] Hardy, Y., "Tunable Lasers," *Photonics Spectra*, Sept. 2000, pp. 133–136.

[70] Commercial Doc. Lucent, "Tunable Laser Module Can Surf the Channels," *Fibre Systems Europe*, Nov. 2000, p. 76.

[71] Hardy, S., "Tunable Lasers Go Electric," *Lightwave*, Nov. 2000, p. 39.

[72] Adams, D. M., et al., "Tunable Laser Co-Packaged with Frequency Locker Delivering over 40 mW of Fiber-Coupled Power on 33 Channels," *ECOC'2000*, Post-deadline papers 3.1, Munich, Sept. 3–5, 2000.

[73] Delorme, F., et al., "Widely Tunable DBR Lasers with Wavelength Drift Lower than 0.009 nm/yr over All Addressable Wavelength Channels," *IOOC-ECOC'97 Proc.*, Vol. 1, No 448, IEE Ed., Edinburgh, Sept. 22–25, 1997, pp. 34–37.

[74] Wei, L. and J. W. Y. Lit, "Er^{3+} Doped Fiber Ring Laser with an External Fiber Bragg Grating," *LEOS'97 Proc.*, Vol. 2, San Francisco, CA, Nov. 10–13, 1997, pp. 382–383.

[75] Kurkov, A. S., et al., "Application of $\pi/2$ Phase Shifted Bragg Grating for the Longitudinal Mode Selection of Single-Frequency Fiber Laser," *IOOC-ECOC'97 Proc.*, Vol. 4, No 448, IEE Ed. Edinburgh, Sept. 22–25, 1997, pp. 49–56.

[76] Vossler, J. L., C. J. Brooks, and K. A. Winick, "Planar Er/Yb Glass Ion-Exchanged Waveguide Laser," *Electron. Lett.*, Vol. 31, No 14, 1995, p.1162.

[77] Hsu, K., et al., "Wavelength Tuning in Efficient Er/Yb Fiber Grating Lasers," *IOOC-ECOC'97 Proc.*, Vol. 4, No 448, IEE Ed., Edinburgh, Sept. 22–25, 1997, pp. 45–48.

[78] An, H. L., et al., "Novel Er-Doped Fiber Loop Laser," *OECC'98 Technical Digest*, Chiba, Japan, July 1998, pp. 304–305.

[79] Barbier, D., et al., "Sub-Centimeter Length Ion-Exchanged Waveguide Lasers in Er/Yb Doped Phosphate Glass," *IOOC-ECOC'97 Proc.*, Vol. 4, No 448, IEE Ed., Edinburgh, Sept. 22–25, 1997, pp. 41–43.

[80] Corning, U.S. Patent, June 1, 1999, W0 00/74186.

[81] Pattison, D. A., et al., "Dual-Wavelength Mode-locked Erbium Fibre Laser with Low Inter-Pulse-Stream Timing Jitter," *ECOC'95 Proc.*, Vol. 1, Brussels, Sept. 17, 1995, pp. 139–142.

[82] Culverhouse, D. O., et al., "Sliding Frequency Er^{3+}/Yb^{3+} Soliton Laser Employing All-Fibre Acoustooptic Frequency Shifting/Tunable Filter," *ECOC'95 Proc.*, Vol. 1, Brussels, Sept. 17, 1995, pp. 143–146.

[83] Richardson, D. J., "Fibre-Based Short Pulse Generation and Shaping Technology," *ECOC'95 Proc.*, Vol. 1, Brussels, Sept. 17, 1995, pp. 147–155.

[84] Chapuran, T. E., et al., "Broadband Multichannel WDM Transmission with Superluminescent Diodes and LEDs," *IEEE Global Telecom Conf., Globecom'91 Proc.*, Vol.1, Phoenix, AZ, Dec. 1991, pp. 612–618.

[85] Lee, J. S., Y. C. Chung, and D. J. DiGiovani, "Spectrum-Sliced Fiber Amplified Light Source for Multichannels WDM Applications," *IEEE Photon. Technol. Lett.*, Vol. 5, 1993, pp. 1458–1461.

[86] Morioka, T., et al., "1 Tbit/s (100 Gb/s × 10 channels) OTDM/WDM Transmission Using a Single Supercontinuum WDM Source," *Electron. Lett.*, Vol. 32, 1996, pp. 906–907.

[87] De Souza, et al., "Wavelength-Division Multiplexing with Femtosecond Pulses," *Opt. Lett.*, Vol. 20, 1995, pp. 1166–1168.

[88] Laude, J. P., et al., "Le Multiplexage de Longueurs d'Onde," *Opto '82 Conf. Proc.*, Paris: Masson Ed., 1982, pp.144–147.

[89] Laude, J. P., "Les Multiplexeurs de Longueurs d'Onde en Télécommunication Optique," *J. Optics*, Vol. 15, No. 6, Paris, 1984, pp. 419–423.

[90] Bouchet, J. M., J. Mayer, and J. P. Laude, "Procédé et Dispositif de Conduite et de Surveillance d'Une Installation Industrielle par Transmission d'Informations et d'Ordres Par Voie Optique," French patent 85 09 335, June 19, 1985, and 85 10 966, July 17, 1985.

[91] Laude, J. P., and J. Lerner, "Wavelength Division Multiplexing/Demultiplexing (WDM) Using Diffraction Gratings," *Proc. SPIE 503*, San Diego, CA, 1984, pp. 22–28.

[92] Reeve, M. H., et al., "LED Spectral Slicing for Single Mode Local Loop Applications," *Electron. Lett.*, Vol. 24, No. 7, March 31, 1988, pp. 389–390.

[93] Hunwicks, A. R., L. Bickers, and P. Rogerson, "A Spectrally Sliced Single-Mode Optical Transmission System Installed in the UK Local Loop Networks," *Globecom 89, IEEE Global Telecommunication Conference and Exhibition*, Vol. 3, Dallas, TX, Nov. 27–30, 1989, pp. 1303–1307, (Pub. IEEE 1989, New York).

[94] Wagner, S. S., and T. E. Chapuran, "Broadband High Density WDM Transmission Using Superluminescent Diodes," *Electron. Lett. (U.K.)*, Vol. 26, No. 11, May 24, 1990, pp. 696–697.

[95] Hunwicks, A. R., "The Effect of Transmitter Wavelength Variations in Spectrally Sliced Optical Transmission Systems," *Proc. SPIE*, Vol. 1179, "Fiber Networking and Telecommunication," 1989, pp. 15–25.

[96] Murtaza, G., "Impact of Thermal Drifts in LED Spectrally Sliced WDM Systems," *Optics Com.*, Vol. 161, March 15, 1999, pp. 318–329.

[97] Arya, V., and I. Jacob, "Interchannel Interference Effects in Spectrum-Sliced WDM Systems," *Proc. SPIE,* Vol. 3491, 1998, pp. 599–604.

[98] Bersiner, L., D. Rund, "Bi-Directional WDM Transmission with Spectrum Sliced LEDs," *Opt. Com.*, Vol. 11-2, 1990, pp. 56–59.

[99] Laude, J. P., F. Bos, and I. Long, "LED Spectral Slicing Wavelength Division Multiplexers for Single Mode Fiber Network," *Opto. '91 Proc,* Paris: Masson Ed., 1991, pp. 397–398.

[100] Lampard, G. J., "Spectrum Slicing of Light Emitting Diodes for Distribution in the Local Loop," *A.T.R.*, Vol. 26, No. 1, 1992, pp. 56–57.

[101] Jung, D. K., et al., "Wavelength-Division-Multiplexed Passive Optical Network Based on Spectrum-Slicing Technique," *IEEE Photon. Technol. Lett.*, Vol. 10, No. 9, Sept. 1998.

[102] Koyama, F., "High Power Superluminescent Diodes for Multi-Wavelength Light Sources" *LEOS'97, IEEE Lasers and Electro-Optics Society Annual Meeting,* Vol. 1, San Francisco, CA, Nov. 1997, pp. 333–334.

[103] Liou, K. -Y., et al., "A 24-Channel WDM Transmitter for Access Networks Using a Loop-Back Spectrally-Sliced Light-Emitting Diode," *IOOC-ECOC'97 Proc.*, Vol. 3, No. 448, IEE Ed., Edinburgh, Sept. 22–25, 1997, pp. 367–370.

[104] Alfano, R. R., and S. L. Shapiro, "Emission in the 4000 to 7000 Ä Via Four-Photon Coupling in Glass," *Phys. Rev. Lett.*, Vol. 24, 1970, p. 584.

[105] Boivin, L., and B. C. Collings, "Spectrum-Slicing of Coherent Sources for Optical Communications," *ECOC'2000 Proc.*, Vol. 3, Munich, Sept. 3–5, 2000, pp. 49–50.

[106] Morioka, T., "Supercontinuum Lightwave Optical Sources for Large Capacity Transmission," *ECOC'95 Proc.*, Vol. 2, Brussels, Sept. 17, 1995, pp. 821–832.

[107] Laude, J. P., *Wavelength Division Multiplexing,* Chapter 15.5, London: Prentice Hall, 1993.

[108] Boivin, L., et al., "110 Channels × 2.35 Gb/s from a Single Femtosecond Laser," *IEEE Photon. Technol. Lett,* Vol. 11, No. 4, April 1999.

[109] Takara, T., et al., "Over 1000 Channel Optical Frequency Chain Generation from a Single Supercontinuum Source with 12.5 GHz Channel Spacing for DWDM and Frequency Standards," *ECOC'2000,* Post-deadline papers 3.1, Munich, Sept. 3–5, 2000.

[110] Campi, D., and C. Coriasso, "Wavelength Conversion Technologies," *Photonic Network Com.*, Vol. 2, No. 1, Jan./March 2000, pp. 85–95.

[111] Karasan, E., and E. Ayanoglu, "Performance of WDM Transport Networks," *IEEE J. on Selected Areas in Comm.*, Vol. 16, No. 7, Sept. 1998, pp. 1081–1096.

[112] Wiesenfeld, J. W., "Wavelength Conversion in WDM Networks," *LEOS'97, IEEE Lasers and Electro-Optics Society Annual Meeting,* Vol. 1, San Francisco, CA, Nov. 1997, pp. 88–89.

[113] Danielson, S. L., B. Mikkelsen, and P. B. Hansen, "Wavelength Conversion and Devices," *LEOS'97, IEEE Lasers and Electro-Optics Society Annual Meeting,* Vol. 1, San Francisco, CA, Nov. 1997, pp. 94–95.

[114] Tucker, R. S., "Technologies for Future Photonic Networks," *OECC'98 Technical Digest,* Chiba, Japan, July 1998, pp. 398–399.

[115] Ramamurthy, B., "Wavelength Conversion in WDM Networking," *IEEE J. on Selected Areas in Comm.*, Vol. 16, No. 7, Sept. 1998, pp. 1061–1073.

[116] Stubkjaer, K. E., et al., "All Optical Wavelength Converters," *OECC'98 Technical Digest,* Chiba, Japan, July 1998, pp. 464–465.

[117] Stubkjaer, K. E., et al., "Wavelength Converter Technology," *IEICE Trans. Electron.*, Vol. E82 C, No. 2, February 1999, pp. 338–348.

[118] Sartorius, B., "Photon Under Control: Optical Signal Processing for High Speed Networks," *IST'2000, Microelectronics, Optronics and Networks Session Proc.*, Nice, France, Nov. 2000, pp. 5–6.

[119] Renaud, M., "Components for Advanced Networks," *OPNET'99 Proc.*, Paris, Jan. 26–28, 1999, pp. 337–348.

[120] Wiesenfeld, J. M., "Wavelength Conversion Using Semiconductor Optical Amplifiers and PICs," *OECC'98 Technical Digest,* Chiba, Japan, July 1998, pp. 456–457.

[121] Owen, M., et al., "All-Optical 1 × 4 Network Switching and Simultaneous Wavelength Conversion Using an Integrated Multi-Wavelength Laser," *ECOC'98 Proc.*, Vol. 1, Madrid, Sept. 20–24, 1998, pp. 649–650.

[122] Kawai, T., et al., "Optoelectronic Wavelength Conversion Scheme Using DFB-LDs and Optical Switch for VWP Photonic Transport System," *OECC'98 Technical Digest,* Chiba, Japan, July 1998, pp. 22–23.

[123] Stephens, M. F. C., et al., "Negative Penalty 10 Gb/s All Optical Wavelength Conversion in an Integrated Amplifier/DFB Laser," *IOOC-ECOC'97 Proc.*, Vol. 2, No 448, IEE Ed., Edinburgh, Sept. 22–25, 1997, pp. 265–268.

[124] Mahgerefteh, D., P. Cho, and J. Goldhar, "Elimination of Pattern Dependance in a Semi-Conductor Optical Amplifier Wavelength Converter Using a Fiber Grating," *IOOC-ECOC'97 Proc.*, Vol. 2, No 448, IEE Ed., Edinburgh, Sept. 22–25, 1997, pp. 273–276.

[125] Mahgerefteh, D., et al., "Enhanced Performance of a Mach-Zehnder Wavelength Converter Using a Fiber Grating," *ECOC'98 Proc.*, Vol. 1, Madrid, Sept. 20–24, 1998, pp. 663–664.

[126] Ueno, Y., et al., "New Wavelength Converter for Picosecond RZ Pulses," *ECOC'98 Proc.*, Vol. 1, Madrid, Sept. 20–24, 1998, pp. 657–658.

[127] Ueno, Y. et al., "168-Gb/s OTDM Wavelength Conversion Using an SMZ-Type All-Optical Switch," *ECOC'2000 Proc.*, Vol. 1, Munich, Sept. 3–5, 2000, pp. 13–14.

[128] Vodjdani, N. et al., "All Optical Wavelength Conversion with SOA's Monolithically Integrated in a Passive Mach-Zehnder Interferometer," *Proc., ECOC'94,* Firenze, Italy, Post deadline paper, Sept. 1994, pp. 95–98.

[129] Schilling, M., et al., "Monolithic Mach-Zehnder Interferometer Based Optical Wavelength Converter Operated at 2.5 Gbit/s with Extinction Ratio Improvement and Low Penalty," *Proc., ECOC'94,* Firenze, Italy, Vol. 2, Sept. 1994, pp. 647–650.

[130] Schilling, M., et al., "Wavelength Conversion Based on Integrated All-Active Three-Port Mach Zehnder Interferometer," *Electron. Lett.*, Vol. 30, Dec. 1994, pp. 2128–2130.

[131] Janz, C., et al., "Mach-Zehnder Wavelength Converter Module with Polarization-Maintaining CW Input Fibre," *IOOC-ECOC'97 Proc.,* Vol. 4, No 448, IEE Ed., Edinburgh, Sept. 22–25, 1997, pp. 192–195.

[132] Mikkelsen, B., et al., "10 Gbit/s Wavelength Converter Realized by Monolithic Integration of Semiconductor Optical Amplifiers and Michelson Interferometer," *Proc., ECOC'94,* Firenze, Italy, Vol. 4, Sept. 1994, pp. 67–70.

[133] Hansen, P. B., et al., "20 Gbit/s Straight Line Cascade of Two Interferometric Wavelength Converters," *OECC'98 Technical Digest,* Chiba, Japan, July 1998, pp. 18–19.

[134] Janz, C., et al., "All-Active Dual-Order Mode Mach-Zehnder Wavelength Converter for Power-Efficient, Co-Propagative Operation," *ECOC'98 Proc.*, Vol. 1, Madrid, Sept. 20–24, 1998, pp. 661–662.

[135] Boffi, P. et al., "Kerr-Based Wavelength Conversion over More Than 30nm," *ECOC'98 Proc.,* Vol. 1, Madrid, Sept. 20–24, 1998, pp. 651–652.

[136] Yu, J. et al., "40-Gb/s All-Optical Wavelength Conversion Based on a Nonlinear Optical Loop Mirror," *J. of Lightwave Technol.*, Vol. 18, No. 7, July 2000, pp. 1001–1006.

[137] Yu, J., et al., "All-Optical Wavelength Conversion of Short Pulses and NRZ Signals Based on a Nonlinear Optical Loop Mirror," *J. of Lightwave Technol.*, Vol. 18, No. 7, July 2000, pp. 1007–1017.

[138] Spiekmann, L. H., et al., "All-Optical Mach-Zehnder Wavelength Converter with Monolithically Integrated DBF Source," *IEEE Photon. Technol. Lett.*, Vol. 9, Oct. 1997, pp. 1349–1351.

[139] Wolfson, D., et al., "Experimental and Theoretical Investigation of the All-Optical Noise Reduction Capabilities of Interferometric Wavelength Converters," *OECC'98 Technical Digest,* Chiba, Japan, July 1998, pp. 20–21.

[140] Inoue, K., et al., "Multichannel Frequency Conversion Experiment Using Fiber Four Wave Mixing," *Electron. Lett.*, Vol. 29, 1993, p.1708.

[141] Grosskopf, G., R., Ludwig, R. Schnabel, and H. G. Weber, "Frequency Conversion with Semiconductor Laser Amplifiers for Coherent Optical Frequency Division Switching," *Proc., of IOOC'89,* paper 19C4-4, Kobe, Japan, July 1989.

[142] Marteli, F., "Semiconductor Optical Amplifiers for Very Efficient Frequency Conversion by Four Wave Mixing," *Proc. IQEC'96,* paper ThN2, Sydney, Australia, July 1996.

[143] Scholz, C. J., L. Dubertrand, and D. J. Blumenthal, "Demonstration of Two Simultaneous Independently Controlled Wavelength Conversions Using a Novel Dual-Pump/Dual-Probe Four Wave Mixing Configuration in a Semiconductor Optical Amplifier," *LEOS'97, IEEE Lasers and Electro-Optics Society Annual Meeting,* Vol. 1, San Francisco, CA, Nov. 1997, pp. 90–91.

[144] Tucker, R. S., et al., "Optical Frequency Conversion by Four Wave Mixing in Semi-Conductor Optical Amplifiers," *LEOS'97, IEEE Lasers and Electro-Optics Society Annual Meeting,* Vol. 1, San Francisco, CA, Nov. 1997, pp. 118–119.

[145] Watanabe, S., et al., "Simultaneous Wavelength Conversion and Optical Phase Conjugation of 200 Gb/s (5 × 40 Gb/s) WDM Signal Using a Highly Nonlinear Fiber Four-Wave Mixer," *IOOC-ECOC'97 Proc.,* Vol. 4, No 448, IEE Ed., Edinburgh, Sept. 22–25, 1997, pp. 1–4.

[146] Watanabe, S., "Broadband Wavelength Conversion and Optical Phase Conjugation Using a Highly-Nonlinear Fiber Four-Wave Mixer," *OECC'98 Technical Digest,* Chiba, Japan, July 1998, pp. 84–85.

[147] Lin, L. Y., J. M. Wiesenfeld, and J. S. Perino, "Polarization-Insensitive Wavelength Conversion Up to 10 Gb/s Using Four-Wave-Mixing in a Semiconductor Optical Amplifier," *LEOS'97, IEEE Lasers and Electro-Optics Society Annual Meeting,* Vol. 1, San Francisco, CA, Nov. 1997, pp. 96–97.

[148] Wiesenfeld, J. M., et al., "Wavelength Conversion Techniques for Optical Networks," *ECOC'98 Proc.,* Vol. 1, Madrid, Sept. 20–24, 1998, pp. 655–656.

[149] Lacey, J. P., R. S. J. Maden, and M. A. Summerfield, "Multichannel Wavelength Conversion Using Four-Wave Mixing," *OECC'98 Technical Digest,* Chiba, Japan, July 1998, pp. 458–459.

[150] Okuno, T., M. Onishi, and M. Nishimura, "Compactly Packaged Wavelength Converter Module Employing Highly Nonlinear Dispersion Shifted Fibers," *OECC'98 Technical Digest,* Chiba, Japan, July 1998, pp. 380–381.

[151] Mecozzi, A., "Frequency Converters Spectral Inverters and Switches Based on Four Wave Mixing in Semiconductor Optical Amplifiers," *LEOS'97, IEEE Lasers and Electro-Optics Society Annual Meeting,* Vol. 1, San Francisco, CA, Nov. 1997, pp. 117–118.

[152] Minch, J., C. S. Chang, and S. L. Chuang, "Wavelength Conversion Using Two Pump Four Wave Mixing in a Double-Moded Distributed Feedback Laser," *LEOS'97, IEEE Lasers and Electro-Optics Society Annual Meeting,* Vol. 1, San Francisco, CA, Nov. 1997, pp. 120–121.

[153] Vahala, K., et al., "TeraHertz Carrier Dynamics and Application to All-Optical Wavelength Conversion," *LEOS'97, IEEE Lasers and Electro-Optics Society Annual Meeting*, Vol. 1, San Francisco, CA, Nov. 1997, pp. 124–125.

[154] Hedekvist, P. O., and P. A. Andrekson, "Fiber-Based Pulsed Wavelength Converter with ± 11nm Range and More than 20 dB of Parametric Gain," *IOOC-ECOC'97 Proc.*, Vol. 2, No 448, IEE Ed., Edinburgh, Sept. 22–25, 1997, pp. 184–187.

[155] Yoo, S. J. B., et al., "Wavelength Conversion by Quasi-Phase-Matched Difference Frequency Generation in AlGaAs Waveguides," *Proc. IEEE/OSA, OFC'95*, Post-deadline paper PD14-2, San Diego, CA, February 1995.

[156] Yoo, S. J. B., "Polarization Independent, Multi-Channel, Multi-Format Wavelength Conversion by Difference-Frequency Generation in Al Ga As Waveguides," *ECOC'98 Proc.*, Vol. 1, Madrid, Sept. 20–24, 1998, pp. 653–654.

[157] Xu, C. Q., H. Okayama, and T. Kamijoh, "LiNbO3 QPM Wavelength Converters for WDM Optical Communication Systems," *OECC'98 Technical Digest*, Chiba, Japan, July 1998, pp. 462–463.

5

WDM and Optical Amplification

5.1 Introduction

At the time of this writing, the solution most frequently used in commercial applications is amplification in erbium-doped single-mode fibers (EDFA). Today, such amplifiers allow error-free DWDM terabit transmissions in the 1,530- to 1,605-nm range over transoceanic distances. However, other rare Earth ions are candidates for WDM and DWDM amplification at other wavelengths and other amplification solutions, such as amplification in semiconductors, amplification through Brillouin and Raman effects in fibers are developed. Particularly, backward Raman amplification distributed along the fiber is often used in hybrid configurations with EDFAs.

5.2 SOAs

5.2.1 Introduction

Much work has been done on SOAs for telecommunication in the past [1–4]. This research field is still very active. In the chapter on active components for DWDM, FP (multi) cavity and other types of semiconductor lasers have been described. Such structures are not adapted to the simultaneous amplification of several wavelengths, but they have been used for tunable wavelength selection. Semiconductor traveling-wave amplifiers (TWAs) allow for amplification over a large wavelength range. These amplifiers have a structure similar to the structure of the FP laser (Figure 5.1), but the

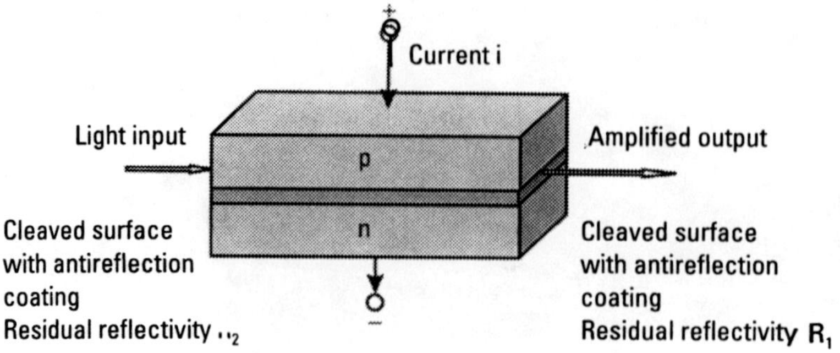

Figure 5.1 Schematic of an SOA.

reflectivity of the cleaved facets is reduced with antireflection coatings. In the Fabry-Perot amplifier (FPA) the facet reflectivities are between one and a few percent, and in a traveling-wave amplifier the reflectivities are as small as possible.

For a semiconductor of length L, with input and output reflectivities at each extremity, R_1 and R_2, respectively, and round trip gain G_a, the power amplification [2, 3, 5] from facet to facet is:

$$A = \frac{(1-R_1)(1-R_2)\exp G_a}{1+R_1\exp(2G_a) - 2R\exp(2G_a) - 2R\exp(G_a)\cos(2\beta L)}$$

with $R = \sqrt{R_1 R_2}$ and $\beta = 2\pi\nu$ n/c (ν is the optical frequency, n the effective refractive index, and c the speed of light). It can be seen that we get a higher optical frequency selectivity with higher reflectivity. In a TWA, the reflectivities are kept small enough (3×10^{-6} in [6] obtained at 1,310 nm with 10°-angle active waveguide with antireflection coatings) and the light to be amplified travels through the device only once, consequently the wavelength range of amplification is wide. The TWA is much more attractive for WDM applications.

5.2.2 FP-Type Amplifiers

As an example of an FPA, using a semiconductor cavity for 1.5 μm with a coated facet at 90% reflectivity, and the other facet without coating, Kazovsky and Werner [5] demonstrated that a 25-dB gain can be obtained with approximately 1.5-GHz spectral width and good adjacent channel suppression (20 dB at 5-GHz frequency offset). The device is tunable over

150 GHz with a change in current of ±10% and a change in temperature of ±7K. The authors predicted the possibility of expanding the tuning range to over 820 GHz with more complex structures (three-section DBR). The device is sensitive to polarization. If the state of polarization of the received signal varies with time, as is generally the case in telecommunications, this problem must be solved.

5.2.3 TWAs

In 1986, Braun, et al. [7] demonstrated that it was possible to use traveling-wave amplifiers for the simultaneous amplification of 10 wavelengths around 830 nm. In 1989, Coquin, et al. [8] showed the simultaneous amplification of 20 wavelengths near 1.54 μm using a TWA with a GaInAsP heterostructure. The active length is 508 μm and an antireflection coating is used on both facets. A 6- to 9-dB gain is obtained if the incident signal level is larger than −32 dBm per channel. In 1990, Koga and Matsumoto [9] demonstrated how the crosstalk depends on the incident power. Typically, they gave a dynamic range of 14.5 dB for 20 channels at 200 Mbp/s with an input power limited to −12 dBm.

5.2.4 Modern Devices

One of the problems with the SOA, when used in WDM amplification, is the nonlinearity and the crosstalk induced by the dependence of the gain of a given channel on the intensity of the other channels. An important improvement has been obtained from gain clamping and from electronic compensation. In gain clamping (B. Bauer, et al. (1994), Keang-Po Ho, et al. (1996), L. J. Tiemeijer, et al. (1996), and others [10]) the amplifier has a feedback on a specific wavelength creating a lasing effect outside of the amplification band. The laser effect clamps the semiconductor carrier concentration. Thus the gain is made independent of the input power. An electronic compensation can be achieved by driving the amplifier with a constant bias plus a signal proportional to the input power in a feed-forward principle. Using this principle, a record total bit rate (160 Gbps) with a SOA gain of 24 dB was demonstrated on a 16-channel WDM transmission with a commercially available SOA [11]. Optical gain up to 30 dB and recovery times as low as 19 ps were obtained [12, 13]. Today, commercially available SOAs find applications mainly because of their affordable cost in medium-distance networks such as access networks in the 1,310-nm window, and as broadband sources for spectral-slicing applications [14].

5.3 Brillouin Scattering Amplifiers

Stimulated Brillouin scattering (SBS) is a nonlinear effect encountered in silica fiber when the power is increased (above a few milliwatts). Frequency-shifted light is generated (mainly Stokes lines at lower frequencies) which provides optical gain that can amplify weak signals. The gain bandwidth is much less than the information bandwidth generally used in telecommunications. However, this effect is very useful for amplification with a narrow spectral selection and has been proposed for amplification and spectral selection. It has been demonstrated [15, 16] that Brillouin scattering amplification allows the selection of channels with separations as low as 1.5 GHz, and a 128-channel, 150-Mbps configuration was tested. However, to our knowledge and up to the time of this writing, SBS amplifiers have not found many applications in practical networks.

A subcarrier multiplexing optical WDM system in which the first-stage selection in WDM is based on Brillouin amplification was analyzed in [17].

5.4 Raman Scattering Amplifiers

Raman scattering, like Brillouin scattering, converts a small portion of an incident frequency into other frequencies. The effect can be stimulated and used in order to transfer energy from a pump laser to a weak signal. In Raman scattering, the Stokes shift is much larger and the gain bandwidth is larger than in Brillouin scattering. Thus, the simultaneous amplification of several multiplexed wavelengths is possible and was proposed in 1987 [18]. However, the problem of crosstalk between channels is particularly difficult to solve [19]. The reduction of the channel interspacing, using a sufficient power and an appropriate pumping wavelength in a "post Raman transmitter fiber Raman amplifier," helped reduce the crosstalk [20–22].

Raman amplification can be operated at any wavelength within the entire window of transparency of silica fibers along the transmission fiber in a distributed scheme. Amplifiers working around 1.3 and 1.5 μm with pumps at 1.24 and 1.42 μm, respectively, in a counter-propagating scheme showed bandwidth of 65 and 100 nm, respectively [22].

Fiber Raman amplifiers are commercialized. They generally require more pump power than EDFAs do. But when used as distributed amplifiers they provide lower nonlinear crosstalks than EDFAs. They can also be used in hybrid configurations with EDFAs to tailor the gain spectrum [23]. In soliton transmission they are well suited for continuous pulse-attenuation compensation along the link.

5.5 Rare Earth-Doped Fiber Optic Amplifiers

5.5.1 Introduction

In these devices, the signal to be amplified and an optical pump are superposed. The pump excites the doping ions: Rare earth such as erbium [24–26], praseodymium, and neodymium [27], to a higher energy level from which amplification takes place by stimulated emission. Among these potential amplifiers, the erbium-fiber amplifiers, suitable for the 1.5-μm transmission window, were extensively studied. Already at *ECOC'92*, R. Laming, et al. and, independently, Lumholt, et al. reported a quantum-noise-limited erbium-doped fiber amplifier with up to 54 dB gain obtained with a special optical isolator between two amplifier stages. So far the erbium-doped fiber amplification represents the maturest industrial choice in DWDM telecommunications. With Er^{3+} the wavelength range is limited to a range of about 75 nm from 1,530 to 1,605 nm on commercially available lasers (about 85 nm at laboratory stage). But other dopants, for example: Nd^{3+}, Pr^{3+} Tm^{3+}, give other wavelengths, namely around 1,300 nm (Pr) [28, 29], from 1,610 to 1,650 nm (Pr) [30], 800 to 830 nm (Tm) [31], and 1,470 to 1,540 nm (Tm) [32]. For instance, Thulium- (Tm) doped fluoride fibers have been used to realize amplifiers in the 800- to 830-nm range with a pump at 785 nm. A gain of 13 dB at 28 mW-launched power was reported [31]. As well Thulium-doped silica fiber amplifiers with a significant gain over the bandwidth 1,470 to 1,540 nm have been demonstrated [32].

5.5.2 Fundamentals of EDFAs

5.5.2.1 Energy Levels

In terms of energy levels, we have a three-level system for the 980-nm pump (Figure 5.2). A two-level system can be used for the pump at 1,480 nm.

In three-level systems, using the pumping photons energy $h\nu_p$ (where ν_p is the optical frequency), the medium reaches the energy state B. By nonradiative energy transfer, the medium is brought to level C with a relatively long relaxation time. The transition from C to ground state corresponds to spontaneous fluorescence. If C is sufficiently well populated, a population inversion between C and A is obtained, allowing stimulated emission from C to A, and hence amplification can take place. Because the ground state is

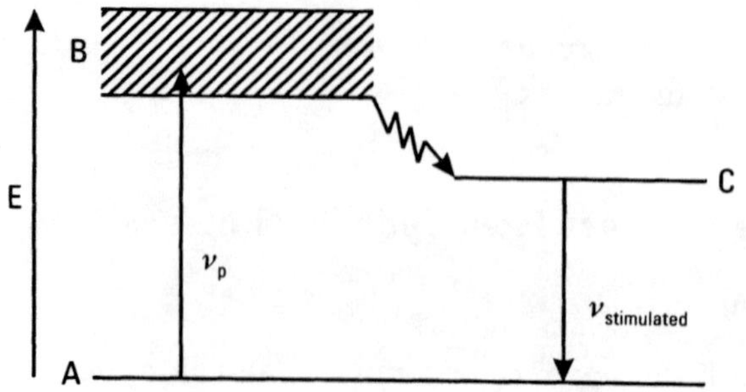

Figure 5.2 Three-level system. Energy levels $E_n\,\nu_p$: pumping, $\nu_{laser} = (E_c - E_A)/h$.

highly populated, the population inversion is generally more difficult to obtain than in four-level systems in which the lower level of the laser transition is not the ground state. However, a complete inversion can be achieved with 980 nm pumping. Maiman gave the equations at equilibrium [33].

With 1,480 nm pumping the population inversion occurs in one step in the $^4I_{13/2}$ level. The quantum efficiency is better than with 980-nm pumping but the noise is larger (Figure 5.3).

5.5.2.2 Host Material

Glass

The host material is generally silica for erbium ions. However, the use of fluoride glasses as host materials has several advantages, among them a lower phonon energy than that of silica, which increases the lifetime at the highest

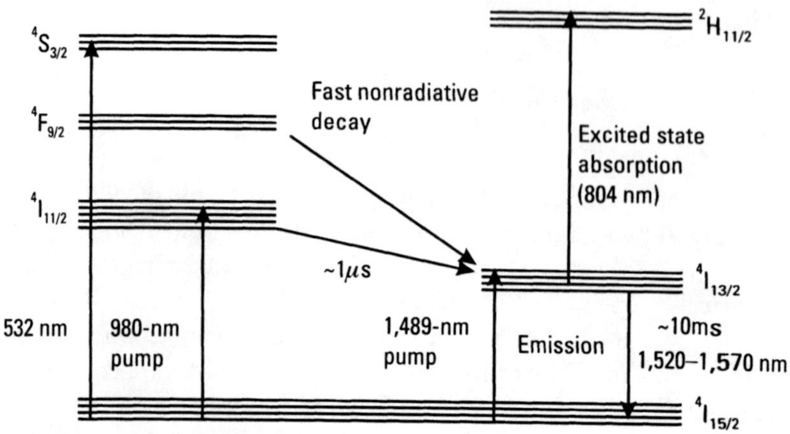

Figure 5.3 Transitions of erbium-doped fiber laser.

metastable levels. However, this prevents pumping at 980 nm and these materials are still difficult to manufacture in volume. So far, ZBLAN fibers are used for praseodymium (the good candidate from the standpoint of maximum gain and amplifying wavelength range at 1.3 μm) [34]).

At approximately 1,600 nm the stimulated emission cross-section of Er^{3+} in telluride glass is twice as large as in fluoride- and silica-based glasses. A bandwidth with 20-dB gain over 80 nm can be obtained from 1,530 to 1,610 nm in telluride glass amplifiers. Furthermore, they can operate up to 1,634 nm, whereas silica and fluoride amplifiers are limited at 1,625 and 1,627 nm, respectively [35].

Codoping

In silica glass and other materials the wavelength range can be extended by codoping with erbium, germanium, phosphorous, and aluminum (Figure 5.4) [36, 37]. In 1989, W. B. Sessa, et al. [38] indicated the possibility of operating alumino-silicate erbium-doped fibers over 30 nm. They were using a fiber doped with 500 ppm of Er^{3+}, on a 1.5-m length with a pumping power of 200 mW at 528 nm. They did not really explore a whole 30-nm range but they came to this conclusion because in their experiment a gain of 22 dB, between the entrance and the exit, was obtained as far as 10 nm outside of the peak gain region. In fact, in 1992, J. Boggis, et al. obtained a maximum gain variation of 2.1 dB over a 30-nm width at about 1,541 nm with an erbium-aluminum codoping [39].

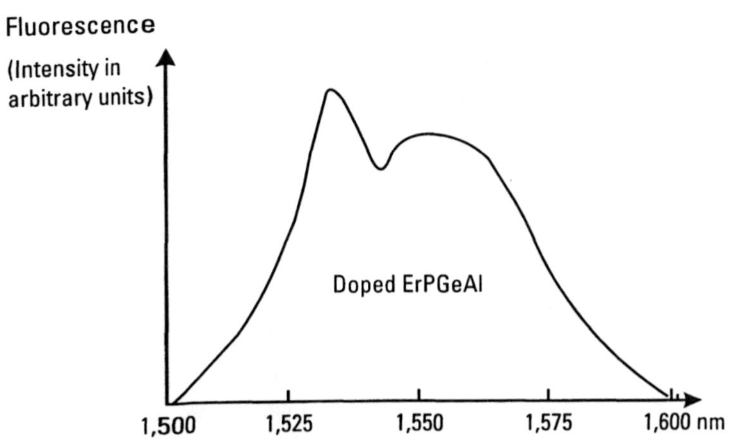

Figure 5.4 A typical ErPGeAl-doping fluorescence spectrum (approximately 20m-long fiber).

More recently, it was shown that a codoping of Er^{3+} and Yb^{3+} gives higher pump absorption and a broader selection of pump wavelength from 800 to 1,100 nm. High output powers could be obtained. For that reason, this solution found applications in CATV in which more than 23 dBm powers are required. For instance, a high output power of 1.2W was obtained at 1,552 nm from a hybrid $Er^{3+}/Er^{3+} - Yb^{3+}$ fiber amplifier with pumps at 980 nm (low-noise preamplifier) and 1,064 nm (high-power section) with a quantum efficiency of 26% [40]. A very-high output power up 16.8W was obtained with a cladding-pumped erbium-ytterbium fiber laser [41].

5.5.2.3 Pump Wavelength

In principle, pumping wavelengths at about 535, 980, or 1,489 nm can be used for erbium [24–26]. Today, EDFAs are usually pumped by semiconductor lasers at 980 and 1,480 nm. In the early days, using a CW argon laser pumped at 514.5 nm and SM silica fibers doped with Er^{3+}, Paynes, et al. obtained lasing with a threshold of 4 mW.

5.5.3 Gain Band

5.5.3.1 Wavelength Range

The gain coefficient depends on the inversion level. In erbium-doped silica fibers with aluminum/germanium codoping, the gain band is flat between 1,540 and 1,560 nm for an inversion level of 40% to 60% [42]. With gain equalization filters, a flat band can be obtained between 1,530 and 1,565 nm in the so C band. In two-stage erbium-doped fiber amplifiers the gain equalizer is generally placed in-between the two stages (Figure 5.5).

Erbium-doped fiber amplifiers operating at longer wavelengths (1,565 to 1,615 nm in the L band and above) have been developed [43–47]. A flat gain can be obtained at lower inversion level of about 40%, thus with a smaller gain coefficient. This means that the doped fiber must be longer, and the pump power must be larger to overcome the loss in this longer fiber. Spontaneous emission is larger and thus the noise figure performance is worse than in C band. For amplification at longer wavelengths (1,570 to 1,630 nm) it was suggested using a pump at 1.53 μm instead of 1.48 μm. This gives twice as much power conversion efficiency [48]. An extension of the useable L band from 1,615 to 1,620 nm was also made possible by compositional modifications to the antimony silicate-glass composition [49]. Of course, erbium fiber-based dual-band fiber amplifiers (DBFA) are proposed with two amplifiers working in parallel to give simultaneously a flat gain in the C and L bands [50].

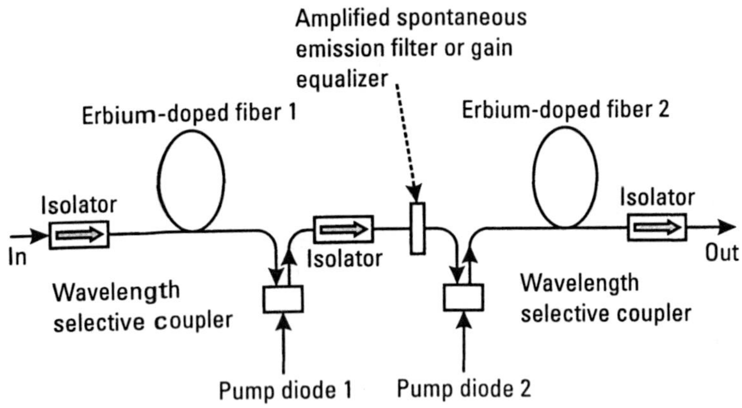

Figure 5.5 A typical two-stage EDFA.

5.5.3.2 Passive or Dynamic Gain Equalization

The gain spectrum of an optical fiber amplifier can be made generically flat in a given relatively small wavelength range as seen above. If a wider bandwidth is required, gain equalization filters are necessary. This is particularly important for long-haul systems containing several cascaded amplifiers, in which a small gain difference within each amplifier would result in large differences among the WDM channels. An early theoretical analysis of the gain and noise spectrum properties of gain-shaped erbium-doped fibers with equalizing filters was given in [51].

A number of approaches to flatten the gain spectrum of erbium-doped fibers have been proposed, including spectral filtering in fiber with novel absorbents or a novel geometry or fiber gratings, acousto-optic devices, and thin film interference filters. We will illustrate these methods with a few examples.

Sm^{3+}-doped fibers were proposed for gain flattening of erbium-doped SM fibers in the range 1.53 to 1.57 μm [52].

Bragg gratings are easy to manufacture and provide one of the most effective ways of gain equalization. Two types of Bragg-grating filters can be used: a cascade of narrow-bandwidth filters on each WDM channels [53, 54], or a wide-band BGF [54].

Gain equalizers made of long-period gratings were used to obtain gain flattening over 84 nm in a C and L band EDFA [55]. A noise figure of 6 dB and an output power of 25 dBm are reported.

S. F. Su, et al. in 1992 [56] already used an acousto-optic tunable filter (AOTF). H. S. Kim, et al. proposed new devices in 1997 [57]. In these filters

Bragg conditions are produced by acoustical waves. The periodical Bragg structures divert partially the light from channels corresponding to their periods. A single-element thin-film-based component that reduces the gain excursion of an EDFA to less than 1 dB from 1,530 to 1,565 nm is presented in [58].

In multichannel networks, fast power transients take place when channels are removed or added. The equalization needs to be dynamically performed, on individual channels or more globally, to combat these effects. Active interferometric filter devices such as MZs, AOTFs, or electronically switchable Bragg gratings (ESBG) can be used to perform a global Fourier-filtering equalization. For example, in the past, a total gain imbalance of 19.3 dB was successfully reduced to 2.3 dB by employing MZ-filter tunable gain equalizers by H. Toba, et al. (*ECOC'92*) on 100-channel FSK signals at 622 Mbps in a 6-cascade in-line amplifier system. Today, ESB gives one of the best response times (50 μs) [59]. However, in a chain of amplifiers a shorter response time of the order of 1 μs for 10 amplifiers will be required [42].

5.5.4 Historical Notes on Amplification in WDM Transmission

At the *OFC'90* conference in San Francisco, California, an experimental demonstration of wideband optical-fiber amplification was reported by Way, et al. using 16 DFB lasers with 2-nm spacing for 6 channels at 622 Mbps and 10 FM TV channels. The same year, M. Hill, et al. [60] built a 10-wavelength, 2-Gbps-per-channel network in the 1,531- to 1,562-nm wavelength range. They showed, with only one of their doped-fiber amplifiers, that it is possible to revolutionize telecommunication networks; for example, sharing a single amplifier between 7,203 customers, to distribute signals at 1.2 Gbps over 30 km. Such solutions led to proposals for broadband networks able to broadcast hundreds of broadband signals to thousands of subscribers [61, 62]. In 1991, ATT Network Service Division conducted a field trial in Roaring Creek, Pennsylvania. Four wavelengths at 17 Gbps were transmitted through the same 520 miles with amplifiers placed 44 miles apart [63]. Moreover, it is worthwhile to note that the fiber amplifier allows the use of the LED-slicing technique (see special section on LED slicing) for distance and bit-rate transmission that can be acceptable in some applications (in 1990, three wavelengths at 1,536, 1,548, 1,560 nm, 140 Mbps and 110 km [64]).

5.5.5 Advantages and Drawbacks of EDFAs

The main advantages and drawbacks of EDFAs are as follows.

5.5.5.1 Advantages

- Commercially available in C band (1,530 to 1,565 nm) and L band (1,560 to 1,605) and up to 84-nm range at the laboratory stage;
- Excellent coupling: The amplifier medium is an SM fiber;
- Insensitivity to light polarization state;
- Low sensitivity to temperature;
- High gain: > 30 dB with gain flatness < ±0.8 dB and < ±0.5 dB in C and L band, respectively, in the scientific literature and in the manufacturer documentation (for instance [65] and EDFA data sheets from the HighWave company);
- Low noise figure: 4.5 to 6 dB (for instance [66] and the HighWave company documentation);
- No distortion at high bit rates;
- Simultaneous amplification of wavelength division multiplexed signals;
- Immunity to crosstalk among wavelength multiplexed channels (to a large extent) [67].

5.5.5.2 Drawbacks

- Pump laser necessary;
- Difficult to integrate with other components;
- Need to use a gain equalizer for multistage amplification;
- Dropping channels can give rise to errors in surviving channels: dynamic control of amplifiers is necessary.

We can see that the advantages overcome the drawbacks, so that frequent and systematic use is likely in multiplexed optical networks.

5.6 Optical Signal-to-Noise Ratio of Erbium-Doped Fiber and Hybrid Raman/Erbium-Doped Fiber Amplifier Transmissions

The amplified spontaneous emission (ASE) of the amplifier degrades the optical signal-to-noise ratio (*OSNR*). The power spectral density at the output of the ASE noise is:

$$P^{\mathrm{ASE}} = 2n_{\mathrm{sp}}(G-1)\,\mathrm{h}\nu \quad \text{(considering both polarizations)}$$

where G is the amplifier gain, ν is the optical frequency, and n_{sp} is the spontaneous emission factor that depends of the inversion population. Its limit is one for a full inversion.

The gain in dB is:

$$G = 10\log\frac{P^{\mathrm{in}}}{P^{\mathrm{out}}}$$

The optical signal-to-noise ratio corresponds to:

$$OSNR = \frac{P^{\mathrm{out}}}{P^{\mathrm{in}}}$$

The noise figure of the amplifier is defined as:

$$N_{\mathrm{F}} = \frac{(S/N)^{\mathrm{in}}}{(S/N)^{\mathrm{out}}}$$

There is a theoretical limit of 3 dB to the noise figure. For commercial amplifiers a typical value of N_{F} is 6/7dB (Figure 5.6).

Unfortunately, on a transmission line each amplifier adds its own noise to the signal. Thus, the optical signal-to-noise ratio of the transmission line depends of the number of optical amplifiers (equivalent to the number of spans N) used along the line. This number is function of the noise figure of the amplifiers.

At the output of a chain of N optical amplifiers, the *OSNR* in dB is given by [68]:

$$OSNR_{\mathrm{G.692}} = P^{\mathrm{out}} - L - N_F - 10\log N - 10\log(\mathrm{h}\nu\Delta\nu_0)$$

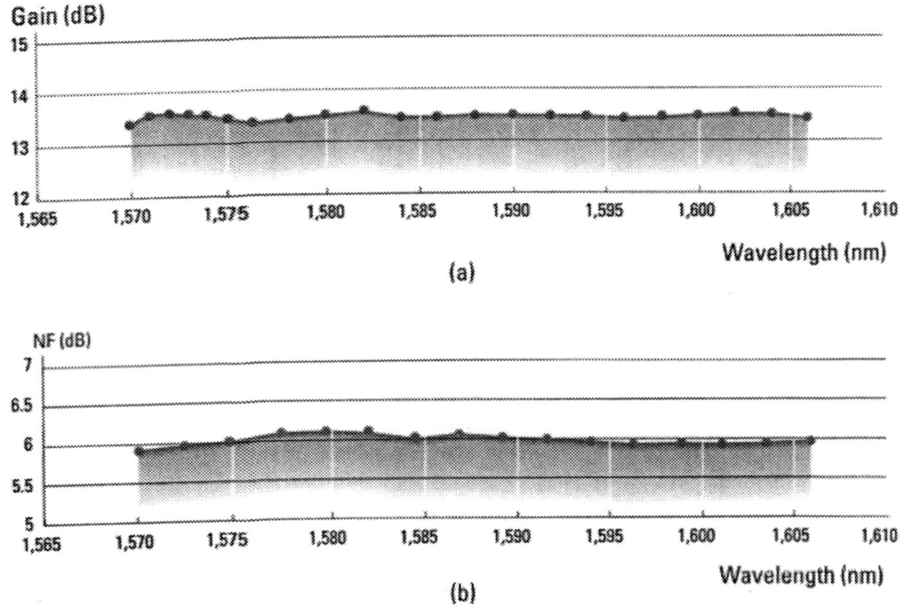

Figure 5.6 Typical gain (a), and noise figure (b) of a L-band booster amplifier. (*Source:* HighWave Optical Technologies. Reprinted with permission.)

Where P^{out} is the channel amplifier output power, L is the span loss between amplifiers, N_F is the noise figure of the amplifiers, hν is the photon energy and $\Delta\nu_0$ is the optical bandwidth.

From a practical measurement point of view, it was shown that this expression agrees completely with the *OSNR* dc measurement performed with an optical spectrum analyzer

$$OSNR_{\text{measured}} = \frac{P}{S_{\text{ASE}} B_0}$$

where P is the average signal power for a WDM channel, B_0 is the optical resolution bandwidth in the optical spectrum analyzer, and S_{ASE} is the spectral density of the accumulated amplified spontaneous emission of the EDFA close to the WDM channel [69].

With an erbium-doped amplifier having typical 4 to 7 dB noise figures at 25-dB gain the signal-to-noise ratio as a function of the number of spans is given in Figure 5.7.

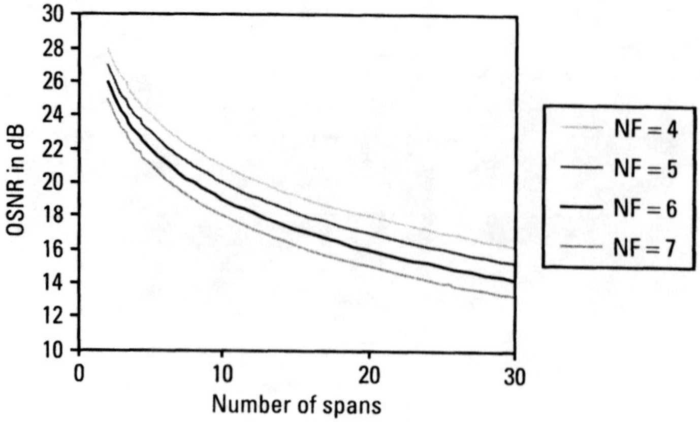

Figure 5.7 Signal-to-noise ratio as a function of the number of spans with erbium-doped amplifiers with 25-dB gain.

In WDM amplification the gain varies with the wavelength. This variation is compensated (see Section 5.5.3.2). In an optical fiber concatenation, the residual difference in gain ΔG between channels produces variations of the *OSNR*. Typically after 10 equally spaced amplifiers, each with $\Delta G = -1.5$ dB, we get $\Delta(OSNR) = -8.7$ dB [70].

With lower gains (shorter spans) the noise is smaller for a given number of spans (Figure 5.8).

Hybrid Raman/erbium-doped fiber amplifiers is a promising new technology for DWDM high bit-rate long-distance links [71, 72]. With an additional Raman amplification (Figure 5.9 and Chapter 7) it is possible to get

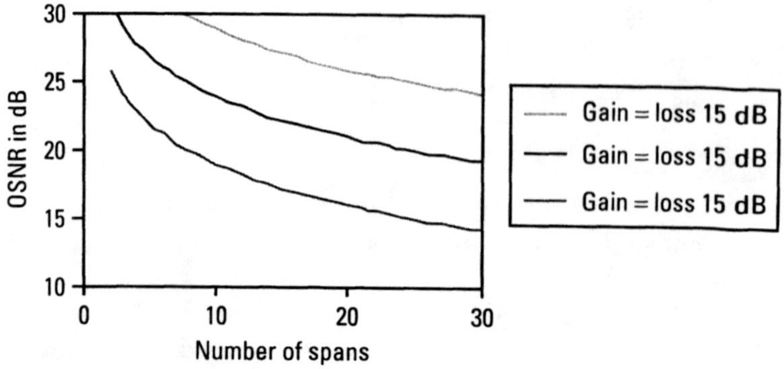

Figure 5.8 Signal-to-noise ratio as a function of the number of spans with erbium-doped amplifiers with different gains but with the same 6-dB noise figure.

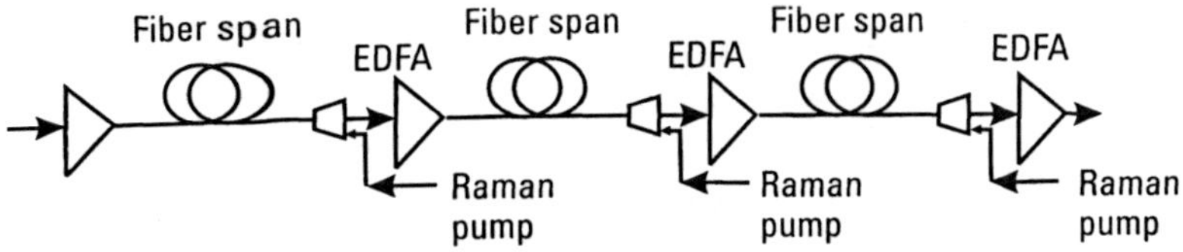

Figure 5.9 Hybrid Raman/erbium-doped fiber amplification.

much better optical signal-to-noise ratios. The equivalent noise figure N_F of Raman amplifiers is negative. For example, a minimum equivalent N_F of −2.9 dB for 40 × 40 Gbps (C band) and 82 × 40 Gbps (C+L bands) WDM channels from 1,530.7 to 1,561.8 nm was obtained with distributed Raman amplification. This gave more than 3-Tbps capacity and 0.4-bps/Hz spectral efficiency [73]. Practical *OSNR* improvements of several dB (or larger fiber spans) can be obtained in long-distance transmissions with hybrid erbium-doped fiber amplifier/Raman amplifier chains [74].

The value of the total N_F in the *OSNR* equation above is reduced. With 6- and 12-dB Raman gains, this lead to the typical curves of Figure 5.10.

5.7 Erbium-Doped Planar Waveguides

These devices, generally called EDWAs, were studied in different places. These compact new devices begin to find applications in metropolitan area networks [75–79].

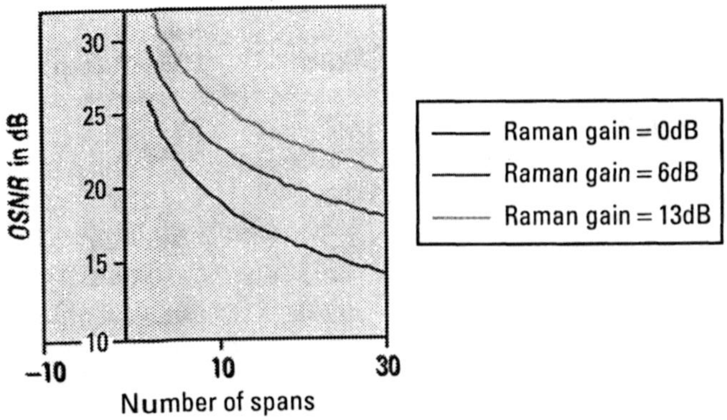

Figure 5.10 Signal-to-noise ratio as a function of the number of spans with erbium-doped amplifiers with 25 gains and 6-dB noise figure and Raman gain. (*After:* K. Hall [74].)

High gain amplifiers have been realized in Er-doped P_2O_5-SiO_2 planar waveguides. 27-dB gain was obtained from a 48-cm amplifier waveguide integrated on a small planar device on silicon of 9 × 5 cm^2 area [75]. A gain of 15 dB at peak wavelength, with 4.5-dB noise figure over the entire C band at 7 dBm output power (up to 12 dBm with a double pumping configuration) was reported [78]. The package size was 130 × 11 × 6 mm. Typical noise factors of 5 dB with 4-dB-per-centimeter gain were reported [79].

5.8 Distributed Optical Amplification

In metro-access network rings, a single laser can be used to optically pump several erbium-doped fiber segments located at discrete points along each ring. A distributed optical fiber amplification (DOFA) takes place at various points of the ring with an advantage of reduced nonlinearities. The location of the pump at the central office facilitates the maintenance [80].

5.9 Comparison of Some Typical Characteristics of the Main Optical Amplifiers

It is rather difficult to compare the characteristics of optical amplifiers. Table 5.1 provides some typical values collected in the literature. In general

Table 5.1
Comparison of Some Typical Characteristics of the Main Optical Amplifiers

	SOA	Raman	EDFA	Hybrid EDFA/Raman	EDWA
Noise figure N_F in dB $(S/N)_{in}/(S/N)_{out}$	6/10	−1/−3 (Negative equivalent N_F)	4/7 Theoretical limit 3 dB	3/4	4.5/7 Theoretical limit 3 dB
Typical gain band nm	80-nm wide	1,300–1,700	1,525–1,620 C + L bands	1,525–1,620 C + L bands	C band, L band feasible
Typical pump power mW	100/300 electrical	150/1,000 optical	10/100 optical	Raman + EDFA pumps	10/15
Typical gain $G = 10 \log (P_{in}/P_{out})$	20/30 dB	6/7 dB per span	10/40	20/40	4 dB per cm 10/15 dB

the SOA amplifiers show larger noise figures but they are cheap, very compact, and they show high extinction rates (above 40 dB) when used as SOA gates. EDFAs are widely used in long-distance transmission because they have high gain and moderate noise figure. EDWAs have less gain but they are more compact and cheaper than EDFAs. With their negative noise figures the Raman amplifiers are now more often used with the EDFAs in long-distance transmission.

Different values may be obtained in specific cases and improvements are announced routinely. Brillouin amplification (unusual) and DOFA (too recent) were not included.

References

[1] Yamamoto, Y., "Characteristics of AlGaAs Fabry-Perot Cavity Type Laser Amplifiers," *IEEE J. Quantum Electronics*, Vol. QE-16, 1980, pp. 1047–1052.

[2] Mukai, T., "Gain, Frequency Bandwidth and Saturation Output Power of AlGaAs DH Laser Amplifiers," *IEEE J. Quantum Electronics*, Vol. QE-17, 1981, pp. 1028–1034.

[3] Simon, J. C., "Semi-Conductor Laser Amplifier for SM Optical Fiber Communications," *J. Opt. Com.*, Vol. 4, Jan. 1983, pp. 51–62.

[4] O'Mahony, M., "Low Reflectivity Semiconductor Laser Amplifier with 20 dB Fibre-to Fibre Gain at 1500 nm," *Electron. Lett.*, Vol. 21, 1985, pp. 501–502.

[5] Kazovsky, L., and J. Werner, "MultiChannel Optical Communications Using Tunable Fabry-Perot Amplifiers," *Appl. Opt.*, Vol. 28, No. 3, 1 Feb. 1989, pp. 553–555.

[6] Tiemeijer, L. F., "High Performance MQW Laser Amplifiers for Transmission Systems Operating in the 1310 nm Window at Bitrates of 10Gb/s and Beyond," *ECOC'95, Technical Digest*, Vol. 1, Brussels, Sept. 17–21, 1995, pp. 259–266.

[7] Braun, R. P., R. Ludwig, and R. Molt, "Ten-Channel Coherent Optic Fibre Transmission Using an Optical Traveling Wave Amplifier," *ECOC'86, Technical Digest*, Vol. 3, Sept. 1986, pp. 29–32.

[8] Coquin, G., et al., "Simultaneous Amplification of 20 Channels in a Multiwavelength Distribution System," *IEEE Photon. Technol. Lett.*, Vol. 1, No. 7, July 1989, pp. 176–178.

[9] Koga, M., and T. Matsumoto, "The Performance of a Traveling Wave Type Semiconductor Laser Amplifier as a Booster in MultiWavelength Simultaneous Amplification," *IEEE J. Lightwave Technol.*, Vol. 8, No. 1, Jan. 1990, pp. 105–113.

[10] Tiemeijer, L. J., et al., "1310-nm DBR-type MQW Gain-Clamped Semiconductor Optical Amplifiers with AM-CATV-grade Linearity," *IEEE Photon. Technol. Lett.*, Vol. 8, 1996, pp. 1453–1455.

[11] Mikhailov, V., and M. Zirngibl, "160 Gbit/s WDM Transmission Using an Electronically-Compensated Semiconductor Optical Amplifier," *ECOC'2000 Proc.*, Vol. 1, Munich, Sept. 3–7, 2000, pp. 79–80.

[12] Dupertuis, M. A., et al., "Fast High Gain and Low-Current SOA by Optical Speed-Up at Transparency," *IEEE Photon. Technol. Lett.*, Vol. 12, No. 11, Nov. 2000, p. 1453.

[13] Bains, S., "SOA Speed-Up Uses Optical Assist," *WDM Solutions*, Feb. 2001, pp. 12–20.

[14] Van den Hoven, G., "Applications of Semiconductor Optical Amplifiers," *ECOC'98 Proc.*, Vol. 2, Madrid, Sept. 20–24, 1998, pp. 5–6.

[15] Chraplyvy, A. R., and R. W. Tkach, "NarrowBand Tunable Optical Filter for Channel Selection in Densely Packed WDM Systems," *Electron. Lett.*, Vol. 22, 1986, pp. 1084–1085.

[16] Tkach, R. W., A. R. Chraplyvy, and R. M. Derosier, "Performance of a WDM Network Based on Stimulated Brillouin Scattering," *IEEE Photon. Technol. Lett.*, Vol. 5, No. 1, May 1989, pp. 111–113.

[17] Lee, Y. H., et al., "Performance Analysis of Wavelength Division and Subcarrier-Multiplexing (WDM-SCM) Transmission Using Brillouin Amplification," *IEE Proceedings J.*, Vol. 139, No. 4, Aug. 1992, pp. 272–279.

[18] Edagawa, N., K. Mochizuki, and Y. Iwamoto, "Simultaneous Amplification of WDM Signals by a Highly Efficient Fiber Raman Amplifier Pumped by High Power Semi-Conductor Lasers," *Electron. Lett.*, Vol. 23, 1987, pp. 196–197.

[19] Weijian, J., and P. Ye, "Crosstalk in Fiber Raman Amplification for WDM Systems," *IEEE J. Lightwave Technol.*, Vol. 7, No. 9, Sept. 1989, pp. 1407–1411.

[20] Kao, M. S., and J. Wu, "Signal Light Amplification by Stimulated Raman Scattering in an N–Channel WDM Optical Fiber Communication System," *IEEE J. Lightwave Technol.*, Vol. 7, No. 9, Sept. 1989, pp. 1290–1299.

[21] Kao, M. S., and J. Wu, "High Density WDM Systems Using Post-Transmitter Fibre Raman Amplifier to Release Raman Crosstalk," *Electron. Lett.*, Vol. 25, No. 21, Oct. 12, 1989, pp.1457–1459.

[22] Chernikov, S. V., et al., "Broadband Silica Fibre Raman Amplifiers at 1.3 μm and 1.5 μm," *ECOC'98 Proc.*, Vol. 1, Madrid, Sept. 20–24, 1998, pp. 49–50.

[23] Masuda, H., et al., "Wide-Band and Low Noise Optical Amplification Using Distributed Raman Amplifiers and Erbium-Doped Fiber Amplifiers," *ECOC'98 Proc.*, Vol. 1, Madrid, Sept. 20–24, 1998, pp. 51–52.

[24] Payne, D. N., et al., "Rare Earth Doped Single Modes Fiber Lasers, Amplifiers and Devices," *Cleo'86 Proc.*, Paper FN1, San Francisco, CA, June 1986.

[25] Mears, R., et al., "Low-Noise Erbium-Doped Fiber Amplifier Operating at 1.54 μm," *Electron. Lett.*, Vol. 23, No. 19, Sept. 10, 1987, pp. 1026–1028.

[26] Desurvire, E., J. Simpson, and P. C. Becker, "High-Gain Erbium-Doped Traveling-Wave Fiber Amplifier," *Opt. Lett.*, Vol. 12, No. 11, Nov. 1987, pp. 888–890.

[27] Mears, R. J., et al., "Neodymium-Doped Silica Single-Mode Fibre Lasers," *Electron. Lett.*, Vol. 21, No. 17, 1985, pp. 738–740.

[28] Carter, S. F., et al., "Quantum Efficiency and Amplification at 1.3 μm in a Pr^{3+} Doped Fluorozirconate Single Mode Fibre," *ECOC'91 Proc.*, Paris, 1991, pp. 21–24.

[29] Yamada, M., et al., "15.1 dB-Gain Pr^{3+} Doped Fluoride Fiber Amplifier Pumped by High Power Laser Diodes Modules," *ECOC'92 Proc.*, Vol. 1, 1992, pp. 49–52.

[30] Choi, Y. G., et al., "Pr^{3+} -Doped Selenide Fiber for 1610–1650 nm Optical Amplifiers," *ECOC'2000 Proc.*, Post-deadline papers 2.1, Munich, Sept. 3–7, 2000.

[31] Percival, R. M., et al., "Amplified First Window Systems Experiments," *ECOC'95 Proc.*, Vol. 2, Brussels, Sept. 17–21, 1995, pp. 913–916.

[32] Cole, B., and M. L. Dennis, "S-Band Amplification in a Thulium Doped Silicate Fiber," *OFC'2001 Proc.*, Paper TUQ3, Anaheim, CA, March 19–21, 2001.

[33] Maiman, T. H., "Stimulated Optical Emission in Fluorescent Solids: Theoretical Considerations," *Phys. Rev.*, Vol. 123, August 15, 1961, pp. 1145–1150.

[34] Miyajima, Y., "Progress Towards a Practical 1.3 μm Optical Fibre Amplifier," *ECOC'92 Proc.*, Vol. 1, 1992, pp. 687–694.

[35] Mori, A., "Telluride-Based EDFAs for Broadband Communication," *OFC'98 Technical Digest*, WA1, Vol. 2, San Jose, CA, 1998, p. 96.

[36] Atkins, C. G., et al., "High Gain Broad Spectral Bandwidth Erbium Doped Fibre Amplifier Pumped Near 1.5 μm," *Electron. Lett.*, Vol. 25. No. 14, 1989, pp. 910–911.

[37] Marcerou, J. F., et al., "Basic Comparison Between Fluoride and Silica Doped Fibre Amplifiers in the 1550 nm Region," *ECOC'92 Proc.*, Vol. 1, 1992, p. 53.

[38] Sessa, W., et al., "Recent Progressing MultiChannel Coherent Lightwave Systems at Bellcore," *Proc. SPIE*, Vol. 1175, 1989, pp. 241–248.

[39] Boggis, J., et al., "Broadband High Sensitivity Erbium Amplifier for Operation over Wide Wavelength Range," *Electron. Lett.*, Vol. 26, No. 8, 1990, pp. 532–533.

[40] Yeniay, A., J.-M. Delavaux, and B. Neyret, "High Power Hybrid Er^{3+}/Er^{3+}-Yb^{3+} Fiber Amplifier," *ECOC'2000 Proc.*, Vol. 2, Munich, Sept. 3–7, 2000, pp. 115–116.

[41] Alan, S. U., P. W. Turner, and A. B. Grudinin, "High-Power Cladding Pumped Erbium-yttErbium Co-Doped Fiber Laser," *OFC'2001 Proc.*, Paper TuI4, Anaheim, CA, March 19–21, 2001.

[42] Sun, Y., et al., "Optical Fiber Amplifiers for WDM Optical Networks," *Bell Labs Technical Journal*, Vol. 4, Jan./March 1999, pp. 187–206.

[43] Massicot, J. F., et al., "High Gain Broadband, 1.6 µm Er^{3+} Doped Silica Fibre Amplifier," *Electron. Lett.*, Vol. 26, No. 20, Sept. 1990, pp. 1645–1655.

[44] Massicot, J. F., R. Wyatt, and B. J. Ainslie, "Low Noise Operation of Er^{3+} Doped Silica Fibre Amplifier Around 1.6µm," *Electron. Lett.*, Vol. 28, No. 20, Sept. 1992, pp. 1924–1925.

[45] Sawada, H., et al., "Broadband and Gain-Flattened Erbium-Doped Fiber Amplifier with + 20 dBm Output Power for 1580 nm Band Amplification," *ECOC'99 Proc.*, Nice, France, Sept. 1999.

[46] Shigematsu, M., et al., "A Novel Configuration of L-Band Erbium-Doped Fiber Amplifier for Improved Efficiency," *ECOC'99 Proc.*, Nice, France, Sept. 1999.

[47] Ono, H., et al., "1.58-µm Band Gain Flattened Erbium-Doped Fiber Amplifiers for WDM Transmission Systems," *IEEE J. Lightwave Technol.*, Vol. 17, No. 3, March 1999, pp. 490–495.

[48] Yamashita, T., et al., "High Efficiency Amplification of EDFA Using a Pump Wavelength of the 1.53 µm Region," *ECOC'2000 Proc.*, Vol. 2, Munich, Sept. 3–7, 2000, pp. 117–118.

[49] Ellison, A. J. G., et al., "Extending the L-Band to 1620 nm Using MCS Fiber," *OFC'2001 Proc.*, Paper TuA2, Anaheim, CA, March 19–21, 2001.

[50] Yang, D., and M. Dignam, "Amplifiers at 1590nm Double the DWDM Bandwidth," *Lightwave Special Report*, August 1999.

[51] Yu, A., M. O'Mahony, and A. S. Siddiqui, "Gain and Noise Spectrum Properties of Gain-shaped Erbium Doped Fibre Amplifiers Using Equalizing Filters," *ECOC'92 Proc.*, Vol. 1, 1992, pp. 481–484.

[52] Belov, A. V., et al., "Sm^{3+}-Doped Fibre Application to the Spectral Filtration in the Range of 1530–1570 nm," *ECOC'92 Proc.*, Vol. 1, 1992, pp. 485–488.

[53] Delevaque, E., et al., "MultiChannel Equalized and Stabilized Gain Amplifier for WDM Transmission," *Electron. Lett.*, Vol. 31, No. 25, 1995, pp. 2149–2150.

[54] Rochette, M., et al., "High-Performance Optical Amplifier for WDM Communication Systems," *SPIE*, Vol. 3491, 1998, pp. 611–616.

[55] Sun, Y., et al., "A Gain-Flattened Ultra Wide Band EDFA for High Capacity WDM Optical Communications Systems," *ECOC'98 Proc.*, Vol. 1, Madrid, Sept. 20–24, 1998, pp. 53–54.

[56] Su, S. F., et al., "Flattening of Erbium-Doped Fibre Amplifier Gain Spectrum Using an Acousto-Optic Tunable Filter," *ECOC'92 Proc.*, WeP 2.3, Vol. 1, 1992, pp. 477–480.

[57] Kim, H. S., et al., "All-Fibers Acousto-Optic Tunable Notch Filter With Electronically Controllable Spectral Profile," *Optical Lett.*, Vol. 22, No. 19, 1997, pp. 1476–1478.

[58] Tilsch, M., et al., "Design and Demonstration of a Thin-Film Based Gain Equalization Filter for C-Band EDFA," *NFOEC'99 Proc.*, Vol. 2, Sept. 26–30, 1999, pp. 390–395.

[59] Ashmead, A., "Spectral Equalization Keeps Optical Signals in Line," *WDM Solutions*, Jan. 2001, pp. 32–38.

[60] Hill, A. M., et al., "7203 User WDM Broadcast Network Employing One Erbium Doped Fibre Power Amplifier," *Electron. Lett.*, Vol. 26, No. 9, 1990, pp. 605–607.

[61] Saifi, M. A., C. Lin, and W. Way, "Optical Fiber Amplifiers for Broadband Optical Network Applications," *FOC LAN 1990,* reprinted in *Fiber Optics*, Jan. 1991, pp. 29–33.

[62] Lin, C., W. Way, and M. A. Saifi, "Optical Fiber Amplifiers Make Broadband Fiber Networks Practical," *Laser Focus World*, Feb. 1991, pp. 161–171.

[63] Phillips Business Information, *Fiber Optics News*, Potomac, MD, July 13, 1992.

[64] Kilkelly, P. D. D., P. J. Chidgey, and G. Hill, "Experimental Demonstration of a Three-Channel WDM System over 110 km Using Superluminescent Diodes," *Electron. Lett.*, Vol. 26, No. 20, Sept. 27, 1990, pp. 1671–1673.

[65] Kimura, Y., M. Nakazawa, and K. Suzuki, "Ultra-Efficient Erbium-Doped Fiber Amplifier," *Appl. Phys. Lett.*, Vol. 57, No. 25, Dec. 17, 1990, pp. 2635–2637.

[66] Shimizu, M., et al., "Erbium-Doped Fibre Amplifier with Extremely High Gain Coefficient of 11.0 dB/mW," *Electron. Lett.*, Vol. 26, No. 20, 1990, pp. 1641–1643.

[67] Zyskind, J. L., et al., "Erbium-Doped Fiber Amplifiers and the Next Generation of Lightwave Systems," *AT&T. Technical J. (U.S.)*, Vol. 71, No. 1, Jan. 1992, pp. 53–62.

[68] *ITU-T Recommendation* G.692, Appendix I, Oct. 1998

[69] Gillner, L., et al., "Experimental Demonstration of the Ambiguous Relation Between Optical Signal to Noise Ration and Bit Error Rate," *NFOEC'1999 Proc.*, Chicago, Sept. 1999, pp. 406–414.

[70] O'Mahony, M. J., "Physical Limitations in Optical Cross-Connect Networks," *ECOC'95 Proc.*, Vol. 4, Brussels, Sep. 17–21, 1995, pp. 63–82.

[71] Masuda, H., et al., "Wide-Band and Low Noise Optical Amplification Using Distributed Raman Amplifiers and Erbium-Doped Fiber Amplifiers," *ECOC'98 Proc.*, Vol. 1, Madrid, Sept. 20–24, 1998, pp. 51–52.

[72] Shimojho, N., et al., "1.22 Tbit/s WDM Transmission Over 7, 221 km with 38nm Bandwidth Expanded by Distributed Raman Amplifier and EDFA," *OFC'2001 Proc.*, Paper WF3, Anaheim, CA, March 19–21, 2001.

[73] Nielsen, T. N., A. J. Stentz, and K. Rottwitt, "Enabling Techniques for DWDM Transmission Systems Having High Spectral Efficiencies," *ECOC'2000 Proc.*, Vol. 3, Munich, Sept. 3–7, 2000, pp. 45–47.

[74] Hall, K., "Next-Generation Optical-Network Designs Overcome Tradeoffs," *Lightwave*, March 2001, pp. 69–78.

[75] Hattori, K., "Er-Doped Planar Waveguide Devices," *LEOS'97, IEEE Lasers and Electro-Optics Society Annual Meeting*, Vol. 2, San Francisco, CA, Nov. 1997, pp. 308–309.

[76] Shmulovich, J., "Erbium Doped Glass Waveguide Amplifier on Silicon," *Photonics West Conf. Proc.*, San Jose, CA, 1997.

[77] Barbier, D., and R. I. Hyde, *Integrated Optical circuits and Components*, Marcel Dekker Inc., 1999.

[78] Barbier, D., "Erbium-Doped Waveguide Amplifiers Promote Optical-Networking Evolution," *Lightwave*, Nov. 2000.

[79] Kilmer, J., "Amplification Advances Show Promise for All-Optical Networks," *Lightwave*, March 2001, pp. 100–106.

[80] Mustafa, D. R., A. G. Abushagur, and B. McNeill, "Building Optical Metro-Access Networks," *Lightwave*, March 2001, pp. 108–114.

6

Routers, Cross-Connects, and Add/Drops

6.1 Introduction

6.1.1 Connections in WDM Networks

WDM networks based on wavelength routing and or add/dropping can offer:

- Optical transparency. This means that a variety of traffic formats and rates can be transmitted on the physical layer without modification to the optical node equipment.
- Smaller processing load and reduced latency in intermediate nodes by bypassing optical traffic not destined to the node at the WDM layer level in both circuit and packet-switch networks.
- Terminal equipment savings as only a subset of wavelengths are necessary at each node.

The overall network is generally described as a set of superimposed layers of connections: physical layer (with fiber sublayer and optical layer), and logical layer (logical connection and logical path). The "virtual" connections between the end systems are supported by a network of logical connections in the logical layer. The connections are one to one, many to one, one to many, or multicast. In a transparent optical network the routing uses wavelength paths. If the wavelengths can be changed along an optical path, the path is called a "virtual" wavelength path.

Designing a full-transparent optical network may imply many unsolved problems. Full transparency means that no digital electronic processing is used and that the network optically performs all filtering, and space and wavelength switching with or without wavelength conversion along all network routes. It can be penalizing to pass through cascades of multiplexers, demultiplexers, passive or active routers, cross-connectors, add/droppers, and different kinds of switches at the different nodes, but moreover, all-optical operation, administration, and maintenance frameworks could be difficult. Then, very often, partial or total optical-electrical-optical conversions are used in the nodes with transponders with or without wavelength translation. In particular, the use of overheads, similar to TDM overheads, and the signal buffering in the optical domain is difficult if not impossible in a fully transparent network. Then the network is often partially "opaque" with some kind of transparency. Electronics are used for complex nonlinear operations but as much as possible the signals are kept in the optical domain to avoid the "147 electronic bottleneck" in high-speed, highly loaded networks. In the long run, all-optical signal processing units (SPUs) will compete with electronic SPUs that are used today for wavelength processing, signal 2R and 3R regeneration, certification of signals accepted on the network, performance assessment before signal delivery, optical channel overhead creation and utilization. Then, very often the nodes use hybrid optoelectronic devices for routers, add/droppers, and cross-connects. However, in this chapter we will present mainly optical routers and cross-connects without discussion of the SPU.

6.1.2 Topological Configurations

6.1.2.1 Introduction

WDM networks generally use star, tree, ring, or mesh topological configurations. Many research projects in different countries have made significant contributions toward creating the future infrastructure based on optical networks using DWDM with time, wavelength, and space switching [1–10].

6.1.2.2 Star Configuration

The star configuration is the usual architecture for the connection between a central computer and its terminals. In telephone networks, this architecture is used for routing bidirectional communications, service, signaling, control, and supervision. In video-communication networks, this topology is generally used between the central office, remote terminals, and customers. In broadband passive optical networks with a multiwavelength star structure, the network terminals are interconnected through a central star coupler. Specific wavelength emitters, as well as specific wavelength receivers, can be allocated to each terminal [11]. The tree configuration is often used between a central computer, data concentrations, and data terminal equipment or, in many cases, in distribution or broadcasting.

6.1.2.3 Ring Configuration

The ring configuration can be used between several data terminals connected to a computer. The FDDI in particular uses this topology, but the ring configuration is used in many other cases.

Interconnected rings

Several rings can be interconnected through optical cross-connects as shown on Figure 6.1. Such networking providing high-capacity, with a flexible, and survivable national-scale multiring network for interexchange and local exchange was demonstrated by the MONET consortium (DARPA funding, U.S.). It was shown to be able of supporting 8-wavelengths, transparently, at 2.5-Gbps modulation rate per wavelength. It used 8-channel wavelength add/droppers and 4×4 wavelength cross connectors (WSXC) [12].

Optical bypass along ring networks with add/drop multiplexers

In WDM rings all wavelengths can be dropped and electronically processed at each node. However, it is generally much better to provide optical bypass at the wavelength layer when possible. This saves electronic equipment and can prevent electronic processing bottlenecks, as it is not necessary to process all traffic entering and leaving each node. For this, a wavelength add/drop multiplexer (WADM) is used in the node. The WADM allows each wavelength to either be dropped and electronically processed or to be bypassed, each wavelength can be added on the ring as well. Traffic grooming can be utilized [13–17] to electronically multiplex lower rate circuits according to

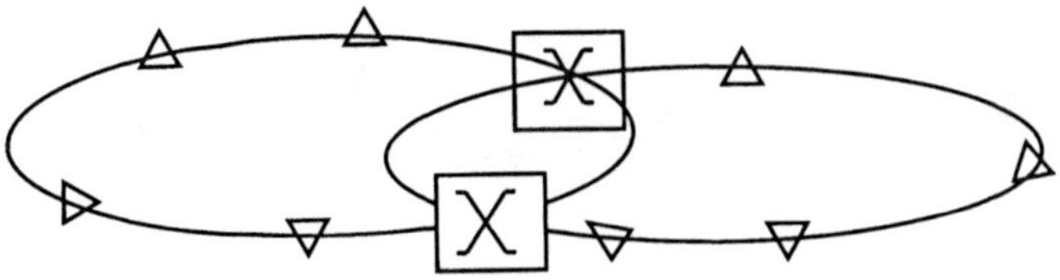

Figure 6.1 Interconnected ring networks.

their destination on dedicated wavelengths, in view of reducing the number of wavelengths that must be electronically processed at each node. In packet-switch networks, a logical topology dynamic reconfiguration using wavelengths may be utilized to reduce the load of the electronic routers. In a few cases of high bit-rate, high-data volumes, an all-optical connection bypassing all electronics in the network may be planned.

In [18] the static circuit assignment problem is discussed. The authors consider a SONET ring using one or two fiber pairs and three wavelengths between four nodes (Figure 6.2). In each node SONET (electronic) add/drop multiplexers (ADM) are used. For example, each SONET multiplexer isable to aggregate 16 OC-3 circuits into an OC-48 (2.5 Gbps). Each wavelength may be used in each node to form three independent OC-48 SONET rings. But, in each node, three expensive ADMs will be necessary. So, 12 ADMs are required in this case. But, as shown on Figure 6.3, it is not necessary to use every wavelength in each node. One wavelength can be bypassed in three of the four nodes. With a circuit assignment of traffic such as:

- Node 1 ↔ Node 2 and Node 1 ↔ Node 3 on wavelength 1;
- Node 2 ↔ Node 3 and Node 2 ↔ Node 4 on wavelength 2;
- Node 1 ↔ Node 4 and Node 3 ↔ Node 4 on wavelength 3.
- Then, only 9, instead of 12, SONET ADMs are necessary.

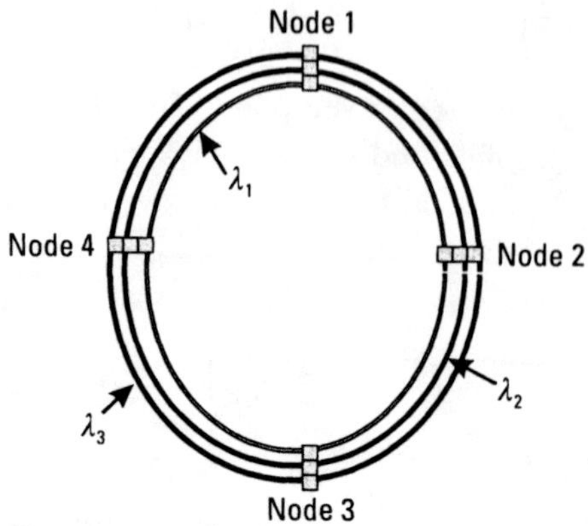

Figure 6.2 Ring network with three wavelengths and four nodes showing an unnecessary use of the three wavelengths at each node (12 SONET ADM).

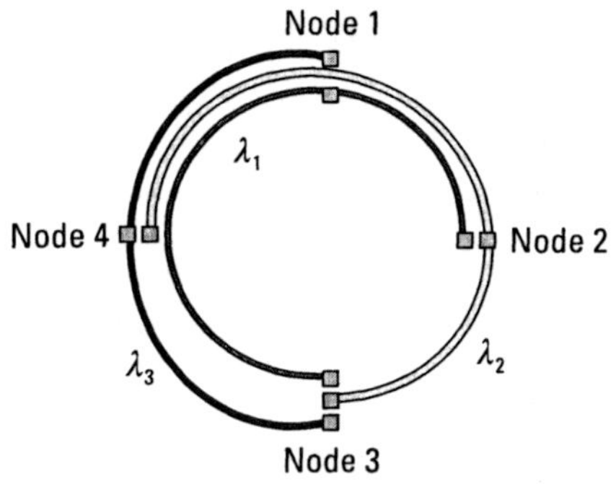

Figure 6.3 Ring network with three wavelengths and four nodes where only two of the three wavelengths need to be used at each node. (9 SONET ADM *After:* [18].)

6.1.2.4 Mesh Configuration

In the mesh configuration (Figure 6.4), more than one path exists for data transmission between two points on the network, with the considerable advantage of reliability and flexibility. The main example is the telephone network.

6.1.3 An Industrial Point of View on IP Transport Networks

In public Internet protocol (IP) backbones, large routers using SONET OC-48c and OC-192c interfaces are deployed. OC192 and OC48 correspond to 10 and 2.5 Gbps, respectively. C stands for "concatenated": all elementary timeslots used in a single application. An optical mesh architecture provides an efficient provisioning of ultra-broadband circuits and an

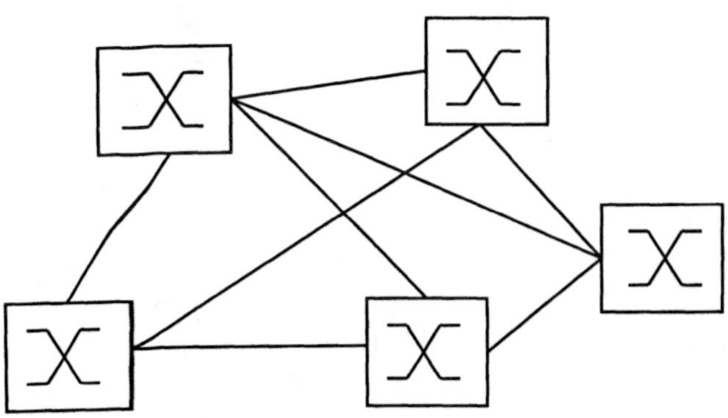

Figure 6.4 Typical mesh network.

optimum use of spare links needed for restoring the traffic outages. The economic advantage of locating the restoration at level 1 (optical level) through optical switches in the optical mesh is explained in [19]. Optical switch "restoration" ports could be substituted for the more expensive IP router ports.

6.1.4 Switching, Routing, and Processing of Signals in the Optical Domain

It is understood that a practical network is very often made up of an association of architectures that constitute the physical medium of the network between stations. In the local network, simple or double star structures are often used. According to the star coupler type used in the central node, the network is called active or passive. The topology is called virtual when it is concerned only with logical connections between stations. One example of an optical multiplexing application is to create virtual topologies on request.

In the network nodes, the function of the switch is to connect any input channel to specified output channel(s) in such a way that the output signals are as close as possible to the input signals. The system allowing this function is a connecting network using time, space, and wavelength switches [20–23].

In space switching, a route is established through an interconnected node network. In the interconnection network, the route may be chosen from many possibilities. Thus, the blocking probability (i. e. the probability that all possible routes are already in use at a given time) can be reduced. The switching principle can be applied to analog or digital transmissions. Time switching is linked to time division multiplexed digital transmissions. In each of the channels of such a multiplex, a time window is allocated: for instance, 3.9 μs in 125 μs cycles for an old standard 30-channel telephone network. The switch modifies the time position of the entrance signal and ensures that the corresponding information train between multiplexes is synchronous. This kind of switching is typically found in ISDN. The space and time techniques perform circuit switching. Moreover, electronic message switching or packet switching allows temporary storage and speedy routing of data messages with their addresses.

In the early 1980s, the great advantages of switching, routing, and processing of signals in the optical domain were increasingly considered. The wavelength dimension could be added. Several groups were pursuing plans to increase the number of multiplexed wavelengths and, should the occasion arise, to use wavelength tunable sources and receivers. A better understanding emerged of the fact that the optical networks could remain totally passive and, for instance, that switching and routing could also use wavelength routers, wavelength-space switches, waveband-space switches, or wavelength-space-time switches in the optical network nodes. This stimulated research on components. It was understood that channel management, provisioning, protection, restoration, routing, and switching of optical channels could migrate progressively from the electronic layer to the optical layer, avoiding optical-to-electronic-to-optical conversions when possible. In some cases, this dramatically reduces the cost of the optical networks. In this chapter, we review the design and performances of DWDM routers, optical cross-connectors, and ADMs enabling interconnection of thousands of inputs with thousands of outputs, with or without tunable sources, and wavelength converters discussed in Chapter 4.

6.2 Wavelength Conversion

6.2.1 Introduction

In all-optical "wavelength continuous" networks, the light-path between two stations is set on a given wavelength that cannot be changed along the route. A new connection requiring a wavelength already allocated to another connection on a given portion of light-path is blocked: Two connections cannot use the same wavelength on the same fiber. This constraint is known as the "wavelength-continuity constraint." Wavelength conversion at the nodes of the network enables the system to avoid this constraint, improving flexibility and efficiency.

If optoelectronic conversion is accepted, the optical signal can be translated in the electronic domain and a tunable laser is set to the new wavelength [24]. For all-optical wavelength conversion, the main technological options are:

- Optical-gating with cross-gain or XPM in semiconductor amplifiers and lasers;
- Wave mixing in a nonlinear medium such as FWM in a SOA or frequency shift in dispersion-shifted doped fibers.

The XPM in an MZ structure appears very promising [25, 26].

6.2.2 Influence of Wavelength Conversion on Network Performances

It is very important to evaluate the potential benefit of wavelength conversion or wavelength interchange within the optical networks through wavelength interchange optical cross-connects (WIXC) [27–35]. No consensus has been reached so far.

Several groups studied the influence of a limited wavelength conversion on WDM network performance under dynamic traffic condition. It was shown that the influence on network blocking limitation could be relatively small above a few wavelength converters per fiber, for Poisson or non-Poisson traffic characteristics, on national-scale long-distance networks with typically 8 wavelengths and 9 nodes [29].

Within the Multiwavelength Optical Networking (MONET) program, several real local-exchange carrier LATA topologies were studied considering the projected traffic requirement using the currently available fiber routes [36]. It was concluded that WDM could "save 25% of the installed equipment costs (with limited configurability) in cases where insufficient fiber is available to support traffic growth or in cases where traffic growth by a factor of five in a few years is needed." For long distances the need of very large WSXC was demonstrated, but with an almost insignificant improvement in throughput by adding wavelength interchange [37].

While there is little difference between wavelength interchange or wavelength selective cross-connection for static traffic demand, it was shown that wavelength interchange improves resource utilization in networks providing restoration or traffic growth capability. The improvement is particularly important in networks with many wavelengths and few fiber links [38].

6.3 Network Architecture Classification

In order to efficiently use the huge bandwidth that optical channels can support, it is necessary to use, in synergy, wavelength, and time and space division multiplexing. In order to establish a circuit connection between two users, a path, time slot(s), and wavelength(s) have to be allocated, avoiding collision (blocking) with other simultaneous connections. The blocking probability is the main measure of quality of service in circuit-switched network [39–41].

The architecture may be purely optical or may include one or more logical network overlays. In the first case, it may use a simple passive broadcast star and a tree topology, or each connection can be routed independently through the network on its own optical frequency through static or dynamic optical routers, WSCX, or wavelength reuse through WICX. In the second case, electronically switched overlays are used on the transparent optical physical network.

The evident explosion in interest in the Internet creates a demand for a considerable increase of capacity and quality in international networking. Much work is devoted to switched DWDM networks with or without wavelength translations. Work is still underway to reach a consensus on what is really required and feasible. Among many other interesting publications see for instance [42, 43].

6.3.1 Broadcast and Select Networks

In many applications, it is possible to use static optical "broadcast and select" networks with switching in the electronic domain only. For instance different stations emit specific wavelengths into a static optical network, through a central star coupler or a bus [44], all optical signals emitted in the networks reach each node where they are optically demultiplexed and electronically switched.

6.3.2 Wavelength Routed Networks

In wavelength routing networks, cross-connection of wavelength division multiplexed signals is obtained through static or dynamic wavelength selectivity in the network nodes. For instance, in a passive star optical fiber network, the $N \times N$ central coupler can be replaced by a wavelength permutation switch, that can be a passive matrix of cross-connections based on WDM [45, 46]. The signal is addressed to a predetermined node, emitted in the network from any given node, on the wavelength corresponding to the destination node. It was shown that for a set of N nodes connected to the central star coupler-multiplexer, it is necessary to have in each node $N - 1$ wavelengths chosen from a set of N-fixed wavelengths.

6.3.3 Linear Lightwave Networks

Linear lightwave networks are wavelength-routed networks with optical nodes using wave-band (with several wavelengths in each wave-band) selective linear divider combiners rather than permutation switches.

6.3.4 Logically Routed Network

On larger networks, when the limits of optics are exceeded, it becomes necessary to add electronic switching in the "logically routed networks." At least one part of the traffic arriving on the node is demultiplexed into small units, electronically processed, routed, converted, and multiplexed again to the output fibers. A logical switching corresponding to a switching at the electronic level helps to keep a full connectivity on complex networks with many nodes in which a fully transparent optical approach is no longer possible or too expensive, requiring too many wavelengths and too many transceivers. SONET networks, ATM networks, and IP networks are typical examples of logically routed networks.

6.3.5 Multigranularity Cross-Connect Architecture

The cross-connect can be made of layers operating at different granularity [47] (Figure 6.5).

- First layer: Fiber cross-connect (FXC) with fibers from and to the network, local fibers added and dropped, and fibers communicating with the wavelength cross-connect layer above.
- Second layer: Wavelength cross-connect (wavelength selective WSXC, or wavelength interchanging WIXC) communicating with the electronic cross-connect layer above. This layer may be divided in two sublayers: waveband cross-connect (WBXC) and wavelength cross-connect.
- Last layer: Electronic cross-connect.

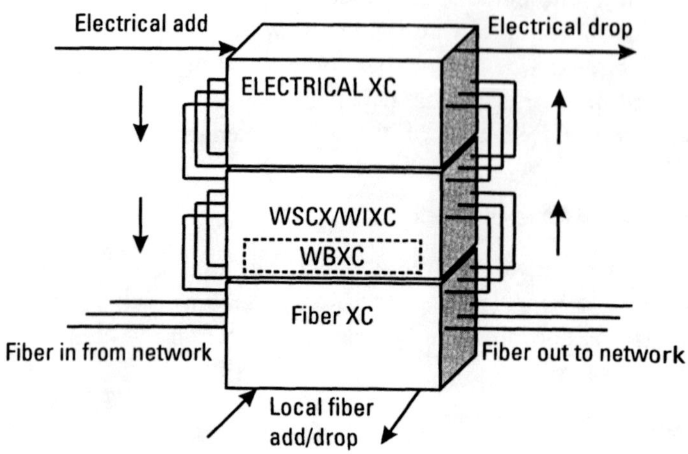

Figure 6.5 Multigranularity cross-connect. (*After:* [47].)

Such a cross-connect can be used in a dynamic grooming architecture in which the best packing of bands of wavelengths onto fibers, the best packing of wavelengths into bands, and the best packing of tributaries onto wavelengths allow an efficient routing in a layer as low as possible, bypassing the layer above without unnecessary demultiplexing and multiplexing.

6.4 Some Definitions Used for Interconnection Performance Characterization

We use the definitions given in [48] for the interconnects.

6.4.1 Bandwidth, Effective Bandwidth, Aggregate Bandwidth

The bandwidth is the maximum number of bits that can be transmitted per unit of time. This is the effective bandwidth of the useful data bits, not including the redundant information (logic-level error correction). The aggregate bandwidth is the total bandwidth of the parallel interconnects: Σ bandwidths (bandwidth of individual interconnect × number of interconnects, for interconnects with identical bandwidths).

6.4.2 Signaling Rate

It is the maximum number of transitions that can take place per unit of time. It is expressed in bauds. Without data encoding (nonreturn-to-zero, NRZ signaling) bandwidth and signaling rate are equal. But if there is clock or low-level redundant information encoded in the transmitted signal, and with other encoding, it is no longer true. For example, with Manchester encoding the peak signaling rate is twice the bandwidth.

6.4.3 Latency

The latency is the time spent to get the electrical information bit input at the sender location (at the output of a standard cell logic gate) available as an electrical information bit output at the receiver end (at a standard logic level). For NRZ, it consists of the delay of the driver circuit, the optoelectronic delay of the light source modulation, the duration of light propagation, and the optoelectronic delay at the receiving end, including the electrical amplifier delay.

6.5 Interoperability in Optical Routed DWDM Networks

The problem is analyzed by Ram Palissery in much detail in [49]. Some of the recommendations corresponding to some specific systems and devices discussed in this chapter are given in the following documents:

- *Draft ITU-T Recommendation G.959.1*, "Optical Networking Physical Layer Interfaces," Geneva, July 1999;
- *GR-1230-CORE*, SONET "Bidirectional Line Switched-Ring Equipment Generic Criteria (a Module of TSGR, FR-440)," Issue 4, Bellcore, December 1998;
- *GR-1312-CORE*, "Generic Requirements for Optical Fiber Amplifiers and Proprietary Dense Wavelength Division Multiplexed Systems," Issue 3, April 1999;
- *GR-1400-CORE*, "SONET Dual-Fed Unidirectional Path Switched ring (UPSR) Equipment Generic Criteria (a Module of TSGR, FR-440)," Issue 2, Bellcore, January 1999;
- *GR-2918-CORE*, "Generic Requirements for DWDM Systems with Digital Tributary for Use in Metropolitan Area Applications: Common Physical Layer Generic Criteria," Issue 3, Bellcore, March 1999;
- *GR-2979-CORE*, "Common Generic Requirements for Optical Add/Drop Multiplexers (OADMs) and Optical Terminal Multiplexers (OTMs)," Issue 2, Bellcore, December 1998;
- *GR-2999-CORE*, "Generic Requirements for WDM Network Management Systems (NMSs)," Issue 1, Bellcore, January 1999;
- *GR-3009-CORE*, "Optical Cross-Connect Generic Requirements," Issue 1, Bellcore, January 1999;
- *GR-2869-CORE*, "Generic Requirements for Operations Based on the Telecommunications Management Network Architecture," Issue 1, Bellcore, October 1995.

6.6 Space Switches

The space-switch cross-connects (or interconnects [50]) fibers connected to the optical node without regard to the attributes of the optical signals. It can be used for traffic routing, reconfiguration, and restoration at the fiber level.

Used with wavelength (or waveband) division demultiplexer/multiplexers and/or static routers it allows wavelength (or waveband) switching.

6.6.1 Crossbar Switches

A fully optical space switch can be realized as a very simple crossbar configuration (Figure 6.6) with three stages that consist of an area of passive splitters, a matrix of controllable on/off switches, and an area of combiners. With N fiber inputs and N fiber outputs it needs N (1 to R) splitters, N × R on-off switches, and R (N to 1) optical combiners. In such a device, the output power is only 1/NR without taking into account any practical losses. High bit-rate hybrid crossbar switches using electrical/optical/electrical (E/O/E) conversion have been proposed.

For example, a cross-bar switch, using the free space optical fan-out of a 8 × 8 VCSEL array, providing 64 replicas of 64-bit serial optical input signals generated by the VCSEL array was reported in [51–54]. A full evaluation of this system is continuing at the time of this writing.

6.6.2 Routers/Selectors Switches and Benes Switches

Large switch matrices can be built using simple y-branched 1 × 2 switches. For example the 4 × 4 switch of Figure 6.7 consists of four switching-unit areas interconnected by three wiring units. This component can be

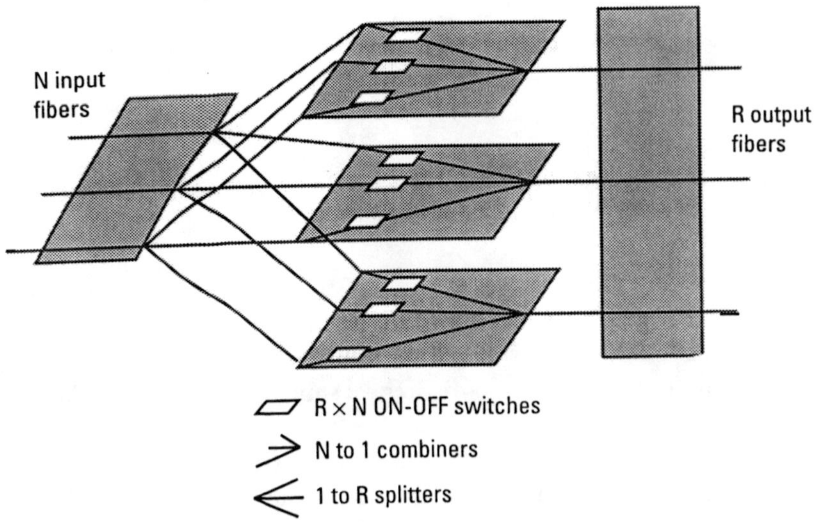

Figure 6.6 Typical crossbar switch architecture.

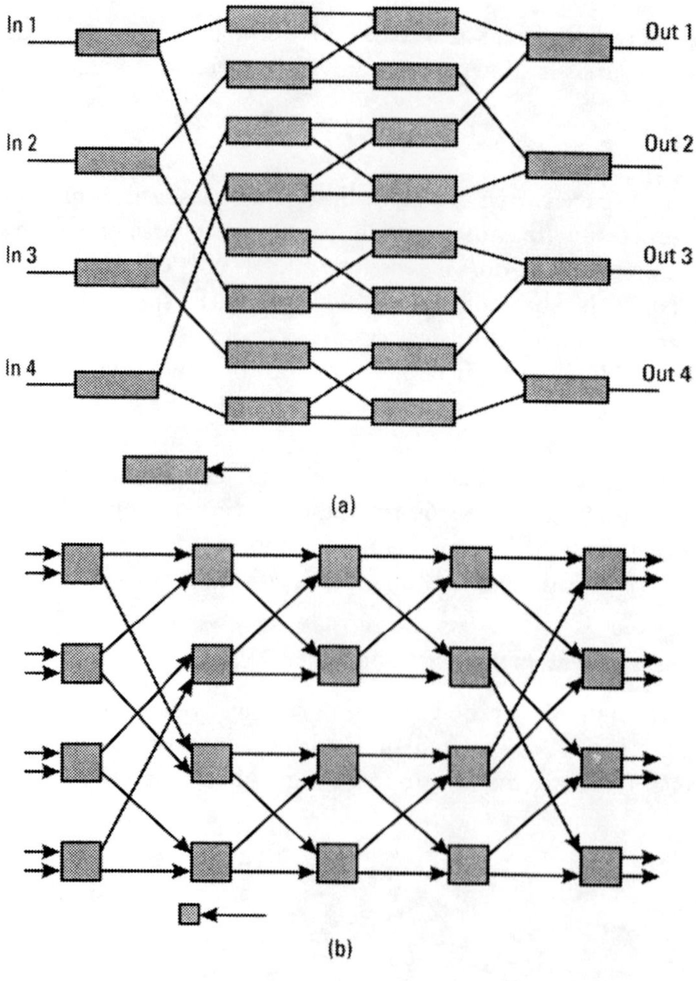

Figure 6.7 4 × 4 router/selector switch and 8 × 8 Benes switch: (a) Y-branched 1 × 2 switch element, and (b) 2 × 2 permutation switch.

integrated and is generally called a router/selector switch. As shown in the figure, a special connection rule allows nonblocking permutation connections. On this device there are no theoretical splitting losses as on the former crossbar device.

The Benes switches use 2 × 2 elementary permutation switches. An N × N router/selector requires 2N (N − 1) elementary switches compared to N^2 for the crossbar and a N × N-dilated Benes architecture switch fabric requires 2N $\log_2 N$ switch elements.

Switch Expansion

A large N × N switch can be made with smaller N/2 × N/2 modules. The design is called switch expansion. In Figure 6.8, an 8 × 8 switch is obtained from four 4 × 4 switches connected through Y couplers. For example, this configuration was used by L. Gillner and G. Gustavson for an amplifier gate switch array [55].

6.6.3 Enabling Technologies

In optical switching several technologies are proposed:

- Optomechanical switching;
- Thermo-optic modifications;
- Liquid-crystal effects;
- Voltage-induced modifications in optical waveguides;
- Cross-gain, XPM in SOAs;
- Acousto-optic effects.

Complementary information is given in [56–58].

6.6.3.1 Optomechanical Switching

Beam Steering

For nonblocking switches with a large number of ports the use of cascades of many 1 × 2 or 2 × 2 elementary switches would give complex

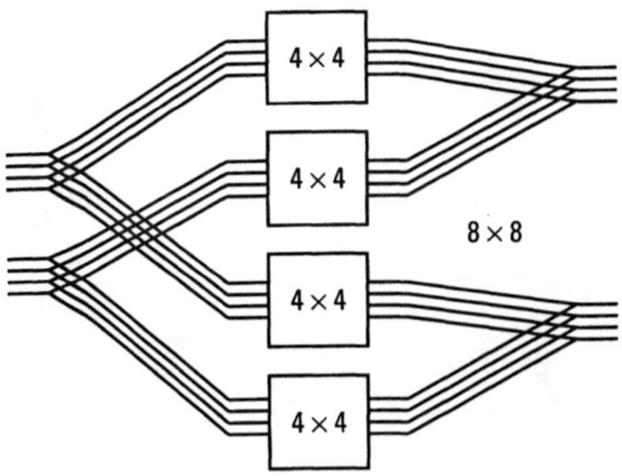

Figure 6.8 Switch expansion: 8 × 8 switch from four 4 × 4 switches. (*After:* [55].)

interconnections and high insertion loss. In beam-steering switches, the complexity of the switch does not increase much with the number of ports.

The configuration of the switch [59, 60] is given in Figure 6.9. Two sets of servo-controlled collimators can be utilized. B. H. Lee and R. J. Capik demonstrated a 576 × 576 device using two-axes positioning of collimators with a total steering angle ±8° with 0.0016° resolution. The insertion loss was 0.7 dB and the isolation 110 dB [59]. In specific cases, phase holograms written on liquid crystals may be utilized to steer the optical beams.

Microelectro-Mechanical Systems (MEMS)

MEMS have progressively emerged as a means of implementing optical mechanical and electrical components on a single wafer [61–68]. It has shown scalability to high port-count and submillisecond switching times [69]. The small mirrors can be 100% reflective or with partial transmission. The losses can be rather small [63, 67]. For example, a free-space micro-mechanical optical cross-connect with bridging functionality for optical-layer restoration developed by L. Y. Lin, et al. [67] shows very-low excess losses: 0.1 and 0.3 dB in the service path and in the restoration path, respectively (Figure 6.10). A free-space add/drop multiplexer with a bulk grating and a micromirror array with good performance has been reported in [65].

6.6.3.2 Thermo-Optic Modifications

Thermo-optic modifications in polymers can be used to realize relatively low-cost optical-switching matrices. For instance, a prototype 8 × 8 non-blocking cross-connect based on polymer switching technology was

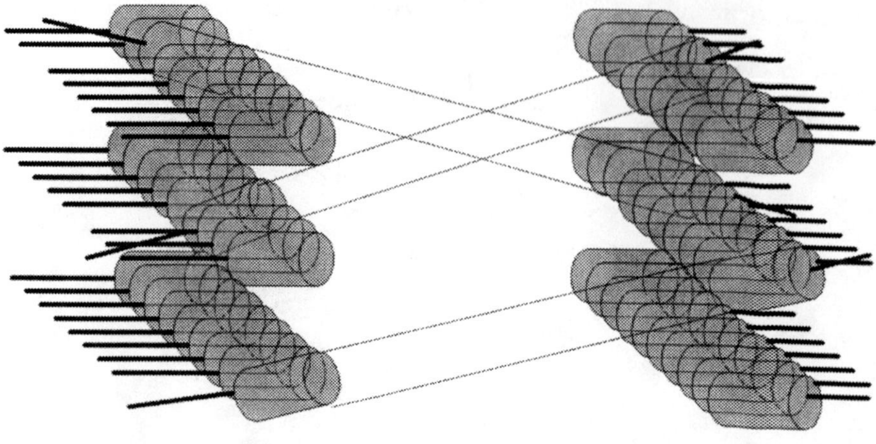

Figure 6.9 Configuration of a beam-steering switch.

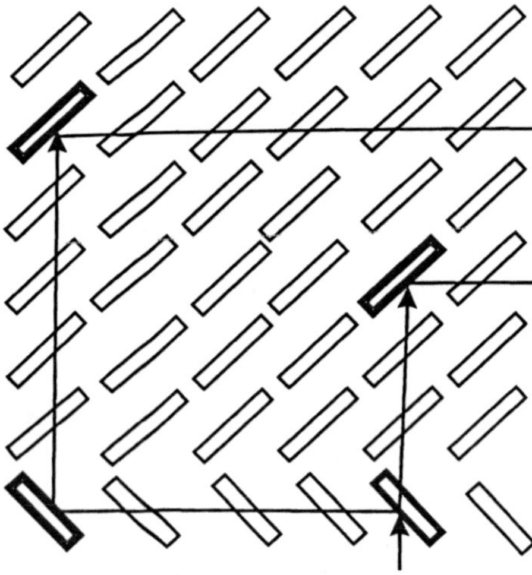

Figure 6.10 MEMS cross-connect for optical-layer restoration of reference. (*After:* [20].)

developed by a large telecommunication component manufacturer [70]. Thermo-optic switches using vertically coupled polymer/silica waveguides combine the advantages of ultra-low-loss silica waveguides and low power consumption of thermo-optic effects [71, 72]. Fluoracrylate composed of pentafluorstyrene, trifluorethylmethacrylate, and glycidylmethacrylate has been used in this application. Acceptable performance on 1 × 2 and 2 × 2 switches (crosstalk −32 and −23 dB in bar and cross-state, respectively, switching power less than 80 mW) was obtained. The crosstalk can be further reduced: for instance, down −42 dB in [73]. Silicone can also be used with good results: extinction ratio up to 30 dB [74], operation even at 80°C.

In [75], results obtained with thermo-optic 8 × 8 matrix switches based on silica-based lightwave circuits are reported. Guided MZ interferometers with thermo-optic phase shift on one arm are used for switching. The average loss and extinction ratio were 5.1 and 60.3 dB, respectively.

Switching using movement by thermo-capillarity of an index matching liquid between two crossed waveguides was also proposed [76].

All these experiments are very interesting (see also [77]), but in our opinion, the practical network applications of thermo-optic switching are still unclear.

6.6.3.3 Liquid-Crystal Effects

Liquid crystal is a material widely used in different commercial applications including high-reliability military applications. For example, commercial liquid-crystal switching elements were tested for over 12,000 hours at 90°C without degradation in optical performance, and optical switch fabrics have been constructed with < 1dB loss and > 50 dB extinction ratios in 1,250- to 1,650-nm wavelength range [78].

Free-Space Switches Using Liquid-Crystal Switching Stages

There are several types of switches based on liquid crystals. They can use reconfigurable gratings based on liquid crystal (Figure 6.11) or polarization manipulation on addressable liquid-crystal arrays placed between birefringent devices. An example of realization of a reconfigurable grating switch is given in [79]. Local spatial frequency holographic recordings on ferroelectric liquid crystal (FLC) are used. An interference pattern can be stored on the FLC by applying a short voltage pulse. A 4 × 1,204 switch was demonstrated. The spatial frequency recorded on this ferroelectric liquid-crystal spatial-light modulator (FLC-SLM) was 37 to 98 lines/mm. It would be a very interesting device for large switching matrices. Unfortunately, the losses are relatively high (28 to 36 dB).

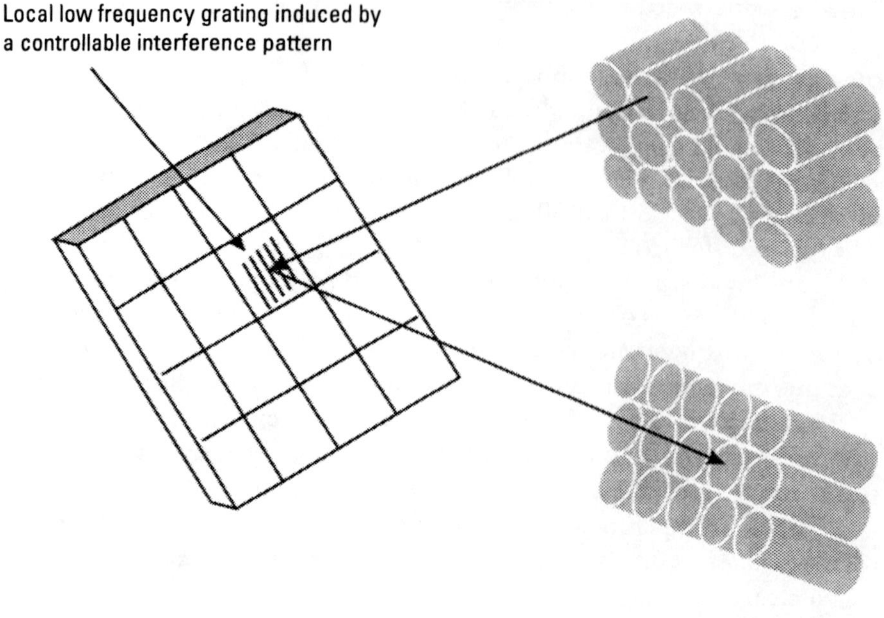

Figure 6.11 Principle of steering in an holographic switch based on ferroelectric liquid-crystal spatial-light modulator. (The collimating devices and the holographic pattern generation means on the reverse side of the FLC-SLM are not shown.)

A different example of reconfigurable grating switch was given in [80]. Here, high-resolution nematic LCs were used, with gratings in tandem (Figure 6.12). An 8×8 switch with about 9-dB loss and an isolation higher than 29 dB was demonstrated. The response time was lower than 50 ms.

In the second type, liquid-crystal free-space switches are made of liquid-crystal transmission-array switches sitting between birefringent spatial-routing stages, arranged between input and output collimating arrays (Figure 6.13). The routing is performed by manipulating the state of polarization of the incoming optical signals in the liquid-crystal pixels placed

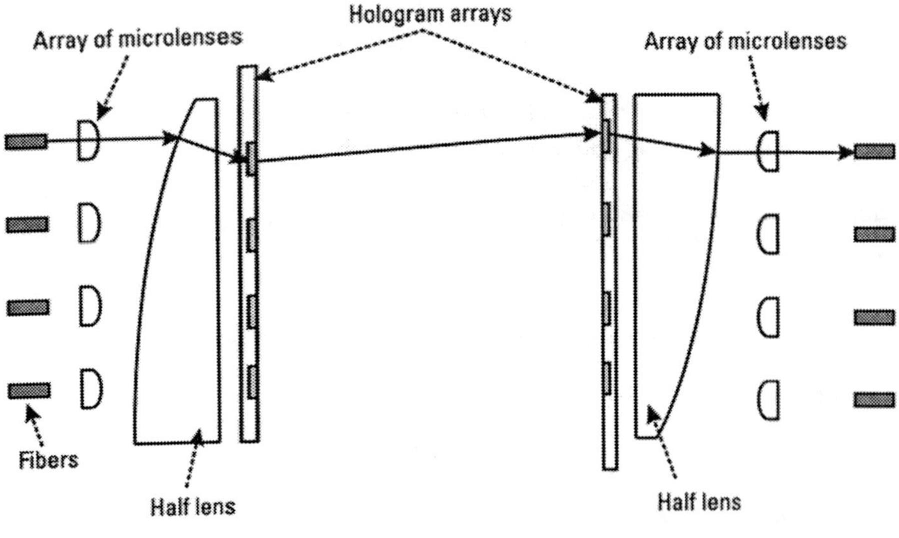

Figure 6.12 Reconfigurable LC grating switch. (*After:* [80].)

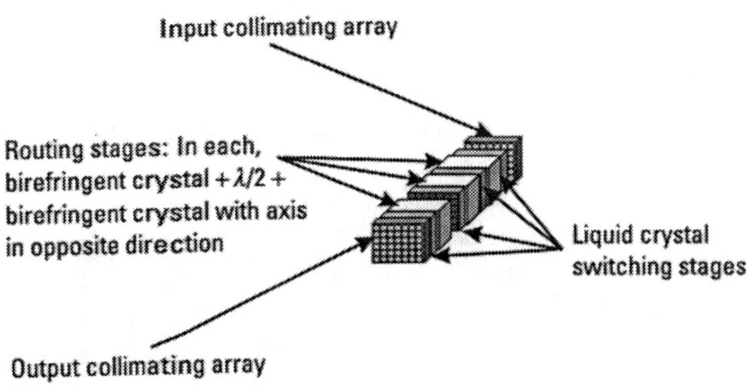

Figure 6.13 Typical liquid-crystal free-space switches.

between the birefringent stages. A change of polarization corresponds to a change in the refraction angle in the birefringent stages.

A 4 × 4 free-space switch for optical cross-connect, designed for a wavelength of 850 nm, was described in [81]. It consists of four polarization control elements, three routing elements, and an optical coupling system. Each polarization control element is a 1 × 4 array of ferroelectric crystal changing polarization state by 90°. Each routing element uses two calcite birefringent crystals, with opposed axis, separated by a $\lambda/2$ wave plate. The array of ferroelectric crystal has four 500 μm × 500 μm cells. The loss was 6 dB and the maximum interchannel crosstalk was –22.8 dB. The response time was 0.2 ms. Cross-connection with 10-Gbps WDM signals using a 64-port liquid-crystal switch was demonstrated in [82] with an insertion loss of 13.2 dB and a crosstalk of about –30 dB.

The performance of an add/drop wavelength-routing switch integrating wavelength filtering and spatial routing with liquid crystal was reported in [83]. 1 × 2, 2 × 2, and 1 × 4 switches using high-speed ferroelectric liquid-crystal shows switching speeds of less than 100 ms and crosstalks better than –40 dB.

6.6.3.4 Voltage-Induced Modifications in Optical Waveguides

Switches on semiconductors

Switch arrays using InP/InGaAsP waveguide Y coupling structures were reported more than 10 years ago. High-speed operation up to 10 GHz was demonstrated in 1995 [84]. These devices can be compact (millimeter size) [85]. Electrorefraction and electroabsorption may be combined to reduce the crosstalk: –40 dB was demonstrated on a 2 × 2-space switch in a dilated Benes structure [86]. For additional information, see also [87].

Switch on lithium niobate

The electro-optic effect in lithium niobate can be used to build many port N × N optical switches with submicrosecond switching time units. The technology is rather mature. However, the thermal sensitivity and the extinction ratios in these devices may sometimes be an issue. An example of realization of an excellent strictly nonblocking 16 × 16 electro-optic photonic switch module was reported in [88]. The device is almost polarization independent. The minimum loss is about 12 dB. The typical extinction ratios are 12 to 17 dB.

6.6.3.5 Cross-Gain, XPM in SOAs

Switch matrices with integrated SOAs have been suggested as they include internal loss compensation. Such 4 × 4 InGaAsP/InP with fiber-to-fiber gain have been realized (Figure 6.14). Spontaneous emission amplification and dynamic gain saturation cause the main limitations. The first effect penalizes the devices at low input levels. The second effect reduces the dynamic range [89–92].

An 8 × 8 SOA gate cross-connector system was developed in 1997 [93] to demonstrate network self-healing. The experiment showed that 20-Gbps signals could be reconfigured in the optical layer within 0.8 ms.

Simultaneous wavelength conversion and routing of 10Gbps optical packets have been demonstrated in a semiconductor vertical coupler optical space switch [94]. The 4 × 4 space switch is fabricated on an InGaAsP/InP wafer. Wavelength conversion of the incoming signals is obtained through XGM of CW sources at determined wavelengths, while the switching is obtained through carrier induced refractive index and gain changes in vertical directional couplers. Outstanding results are also reported in [95–97]. Such devices seem promising for future wavelength/packet switched OXCs.

6.6.3.6 Acousto-Optics Switching

The tunable acousto-optic filter was proposed by Harris and Wallace in 1969 (Figure 6.15) [98]. In the initial experiment, a $LiNbO_3$ crystal was set between a crossed polarizer and analyzer. The incident electric field after transmittance through the polarizer is parallel to the crystal optical axis z and propagates at an extraordinary index. Without the acoustic wave, there is

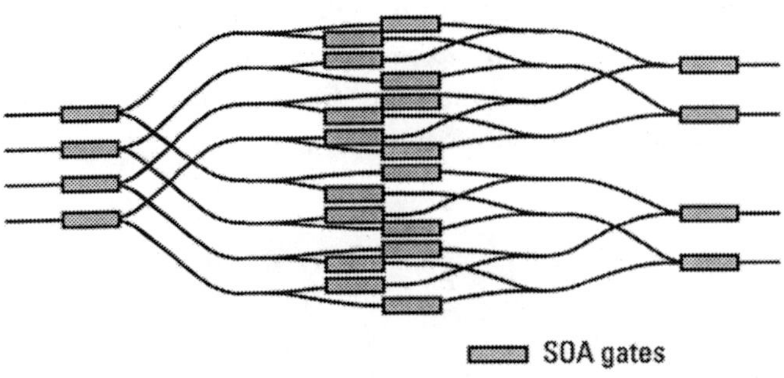

Figure 6.14 Typical layout of an integrated 4 × 4 SOA gate switch matrix.

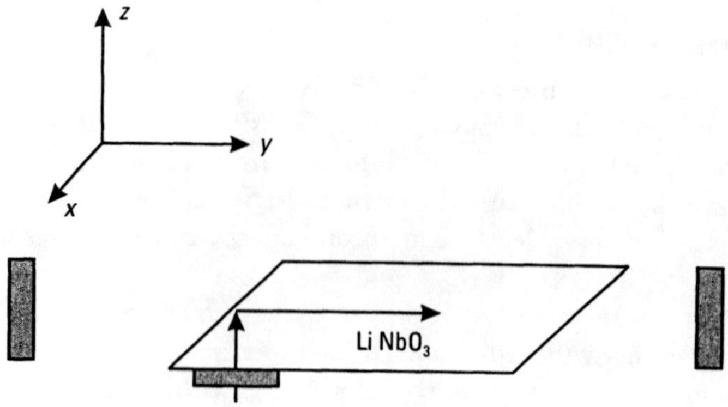

Figure 6.15 Harris and Wallace acousto-optic filter. (*After:* [98].)

extinction through the analyzer at the crystal exit. When an acoustic wave is applied in A and directed in the light-propagation direction *y*, an interaction between the light and the acoustic wave takes place in the crystal.

ω_{ex}, ω_{or} and ω_a being, respectively, the incident optical angular velocity ("ex" for extraordinary), the emerging optical angular velocity "or" for ordinary) and the angular velocity (the angular velocity is related to the frequency ν by $\nu = \omega/2\pi$, we obtain:

$$\text{Incident optical wave:} \quad \hat{E}z(yt) = \left[Ez(y)/2\right] \exp j(\omega_{ex} t + k_{ex} y)$$

$$\text{Acoustic wave:} \quad \hat{S}(y,t) = \left[\frac{S(y)}{2}\right] \exp j(\omega_a t - k_a y)$$

The acoustic wave interacts with the incident optical wave to produce forced optical waves at pulsations $\omega_{ex} + \omega_a$ and $\omega_{ex} - \omega_a$. In lithium niobate, $n_{or} > n_{ex}$, in case of phase matching such that $k_{ex} + k_a = k_{or}$, additive diffraction effects will be obtained such that:

$$\hat{E}x(y,t) = \left[\frac{Ex(y)}{2}\right] \exp j(\omega_{or} t + k_{or} y)$$

This corresponds to the cross-polarization state and therefore, light passes through the analyzer if the wavelength corresponds to $k_{ex} + k_a = k_{or}$, or $\lambda = V_s(n_{or} - n_{ex})/f_0$ where f_0 is the acoustic frequency and V_s the acoustic velocity in the substrate; λ is tunable with f_0. At the beginning, such filters were manufactured with small crystals and, later, on integrated optic devices. The spectral width is such that:

$$\Delta\lambda \approx 0.8 \frac{\lambda^2}{L(n_{or} - n_{ex})}$$

Integrated filters with an interaction length L = 4 mm with $\Delta\lambda$ = 2.5 nm were obtained [99], as well as with $\Delta\lambda$ = 1 nm [100, 101]. Applications of acousto-optic tunable filters were proposed for wavelength selective circuit switching and packet switching [101]. (A few other references are given in our reference section [102–109].)

It was shown that integrated devices in $LiNbO_3$ could be used to design multiwavelength large tuning range switches [110–111]. However, these devices, in general, have much crosstalk and high PDL. The crosstalk can be reduced by dilatation [112]. An example of the realization of an add/drop multiplexer for wavelengths ranging from 1,530 to 1,570 nm is given in [113]. The unpolarized loss is 4.5 dB, the PDL is 2.3 dB and crosstalk is –25 dB.

6.7 Passive Wavelength Router

In a node with input ports (1, 2, 3, …i, …N) and output ports (1, 2, 3, …, j, …M), passive components allowing a wavelength selective cross-connection from N input ports to M outputs can be utilized. These components can connect any input i to any output j at a given wavelength λ_{ij} [114-117].

6.7.1 Node with Interconnected Demultiplexer/Multiplexers

The cross-connection can be made with WDM demultipler/multiplexers utilized in a back-to-back configuration as shown in Figure 6.16. In such a node it is possible to interchange wavelengths between any input and any output fibers. For instance, with λ_N input port 1 is connected to output port N. As well with the same wavelength λ_N the input port N can be connected to the output port 1 without collision. The optical path is determined uniquely by its input port and by its wavelength: A signal arriving on input i, carried on wavelength λ_k, is routed to outport j where

$$k = i + j - 1 \text{ for } i + j \leq N + 1 \text{ and } k = i + j - 1 - N \text{ for } i + j \geq N + 2$$

See Table 6.1.

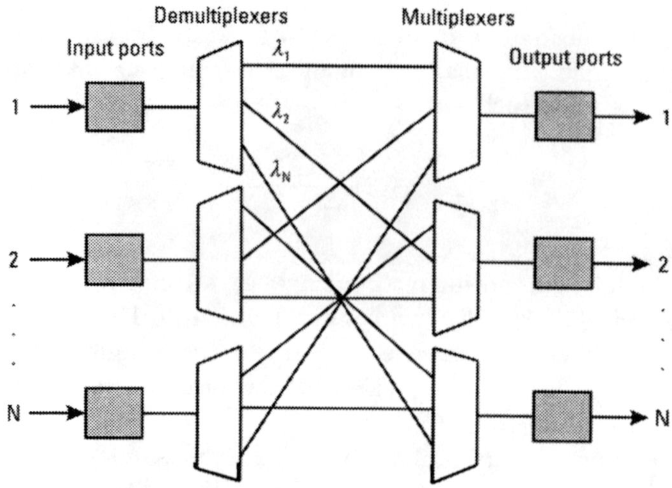

Figure 6.16 Typical WDM cross-connection made with back-to-back demultiplexers/multiplexer.

Table 6.1
Four Wavelengths Used to Connect Any Input Port 1 to 4 to Any Output Port 1 to 4

Input/Output	1	2	3	4
1	λ_1	λ_2	λ_3	λ_4
2	λ_2	λ_3	λ_4	λ_1
3	λ_3	λ_4	λ_1	λ_2
4	λ_4	λ_1	λ_2	λ_3

Here, the routing is fixed without rearrangeability, but we will see later how the addition of a space-division switch allows rearrangeability. We will also see in Section 6.7.2 that passive grating routers, with analogous connecting functions, but avoiding the double losses of demultiplexing before multiplexing of these structures, can be designed in most cases.

The wavelength-routed 4×4 star in Figure 6.17 is a particular case of the back-to-back demultiplexer/multiplexer devices. This node is made of eight identical three-wavelength WDMs interconnected as shown. It allows the design of survivable WDM star networks. The connections from four transmitters, T_1 to T_4, to 4 receivers R_1 to R_4 can be obtained using only three wavelengths as given in Table 6.2 [118].

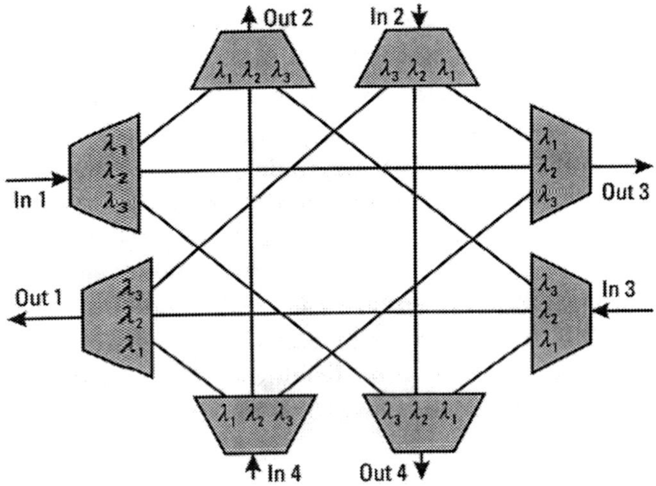

Figure 6.17 Wavelength-routed star. (*After:* [118].)

Table 6.2
Wavelength Routed 4 × 4 Star, Wavelengths Used to Connect Any Transmitter T_1 to T_4 to Any Receiver R_1 to R_4

	R1	R2	R3	R4
T1	—	λ_1	λ_2	λ_3
T2	λ_3	—	λ_1	λ_2
T3	λ_2	λ_3	—	λ_1
T4	λ_1	λ_2	λ_3	—

Remark: Use of multiple multiplexers [119].

It is worth underlining that the eight WDMs of Figure 6.17 can be replaced by a single multiple multiplexer using only two dielectric filters or by a grating multiple multiplexer obtained with a bulk grating or an AWG with 32 fibers connected as explained in [119]. Of course, more generally, a single multiple multiplexer using various wavelength selective functions can always be used to realize the design showed in Figure 6.16.

A coupling function or a wavelength multiplexing function through an optical system, with two or three dimensions, generally sets a relation between a transmission fiber and an entrance or exit fiber set. Thus, for instance, in a Y-coupler, the transmission-line fiber is imaged onto two other fibers through an optical divider (semitransparent mirror or pupil divider) and one or more focusing optics. In a wavelength division multiplexer, the transmission fiber is imaged according to the wavelength at different focal plane locations with a grating (arrayed waveguide grating in AWG or diffraction grating in three-dimensional configuration) and/or a set of multidielectric filters or other wavelength separating means; these locations are associated with one or several focusing optics (except in the concave grating case which combines dispersion and focusing functions). As a matter of fact, all these devices provide a relation between object and image fields (in so far as the stigmatism is sufficient and the relative fiber positions in the different fields are controlled, one could realize several couplers and/or several elementary multiplexers, through the same dividing and focusing optics). In such a component, one coupling optic is used on a set of p fibers arranged in subsets of n + 1 fibers.

6.7.2 Static Grating Routers

6.7.2.1 Principle

A single grating component can connect n inputs to n outputs, each connection from input i to output j requiring a given wavelength λ_{ij} (Figure 6.18). For instance, in a diffraction grating configuration working in a fixed diffraction order, this wavelength is $\lambda_{ij} \approx \lambda_0 + (i + j)\,\Delta\lambda$, where $\Delta\lambda$ is the channel-spacing constant and λ_0 is a constant wavelength (see Section 6.7.2.3). A complete nonblocking connectivity between every input and output port is

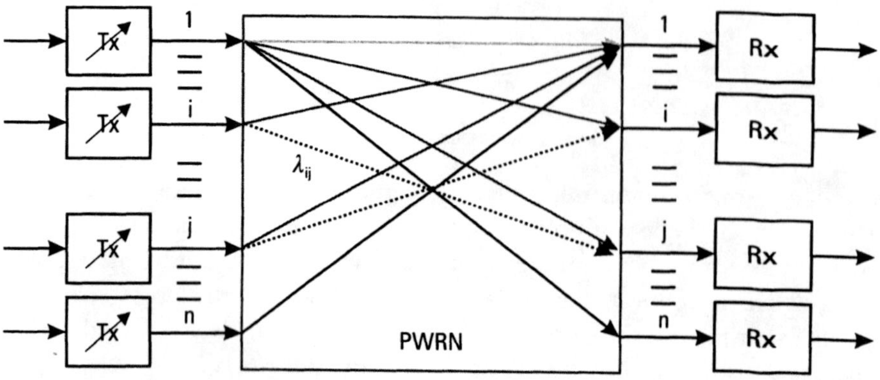

Figure 6.18 Passive wavelength routing node with a diffraction grating configuration working in a fixed diffraction order: $\lambda_{ij} \approx \lambda_0 + (i + j)\,\Delta\lambda$. Using two different diffraction orders, a cyclic permutation of wavelengths can be obtained: $\lambda_{ij} \approx \lambda_0 + (i + j)\,\Delta\lambda$ when $i + j \leq N + 1$, and $\lambda_{ij} \approx \lambda_0 + (i + j - N)\,\Delta\lambda$ when $i + j \geq N + 2$.

provided. Nonblocking in that case means that two connections from two different inputs to the same output cannot be obtained through the same wavelength. Passive optical network architectures based on such components were designed to establish virtual topologies (see, for example, [120]).

6.7.2.2 Typical Applications

This static router is more faults-tolerant than the cross-connectors that also use active devices such as space-division switches. However, the routing of light-paths between fibers through a static router depends only on the wavelength corresponding to a reduced freedom in rearrangeability.

Waveguide-grating routers have been used in the design of access networks [121–123]. A simplified example of a network using static routers is given in Figure 6.19. The 3 × 3 static routers are located in each node of a bidirectional ring connecting three nodes. The connections are AB, AC, BA, BC, CA, and CB (also superfluous connections AA, BB, and CC) [120].

Networks with switching functionality relegated to the edges can be designed with passive components with a high interest for metropolitan area networks (MANs) and wide area networks (WANs). A star network interconnecting N major switch nodes in a national-scale telecommunication network is given in [124].

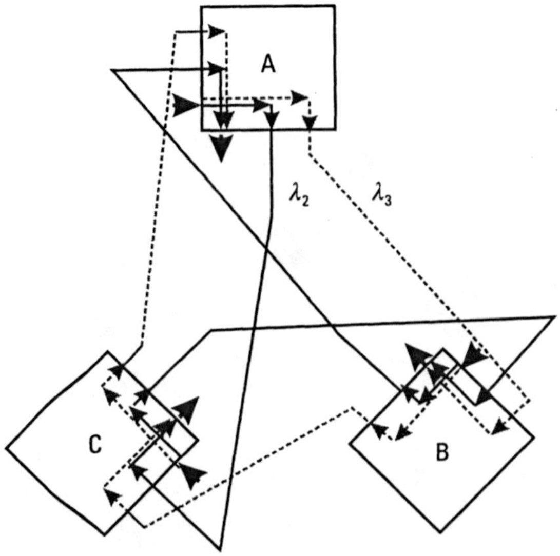

Figure 6.19 A typical application of 3 × 3 static routers in a bidirectional ring connecting three nodes. In each node a three-wavelength laser array and a three-wavelength receiver array are used. (*After:* [120].)

A metro-scale full-mesh network with a single-star topology with 1,024 paths at 10 Gbps each with a 32 × 32 AWG (aggregate capacity 10 Tbps, scalability to 92 Tbps, 9,200 paths) on a 10-km radius was demonstrated in September 2000 [125].

These routers can also be used in particular configurations, for example, to realize optical amplification on bidirectional links. In [126] the authors showed how to amplify 16 channels along a bidirectional link with 8 channels in each direction. The component made of two symmetric router groups all upstream and downstream channels, in the same direction inside the same amplifier, then reinject them in their initial direction in the bidirectional transmission fiber.

6.7.2.3 Static Wavelength Router Based on "Free Space" Diffraction Grating Configuration

Principle

Dense and very dense N*N wavelength routers based on diffraction grating technology are currently available (typical example in Figure 6.20). These devices make use of highly dispersive planes [127–129] or concave [130, 131] gratings to spatially position and separate discrete wavelengths having subnanometer spacing, on a double fiber array figure (Figure 6.21) or even a single fiber array (Figure 6.22).

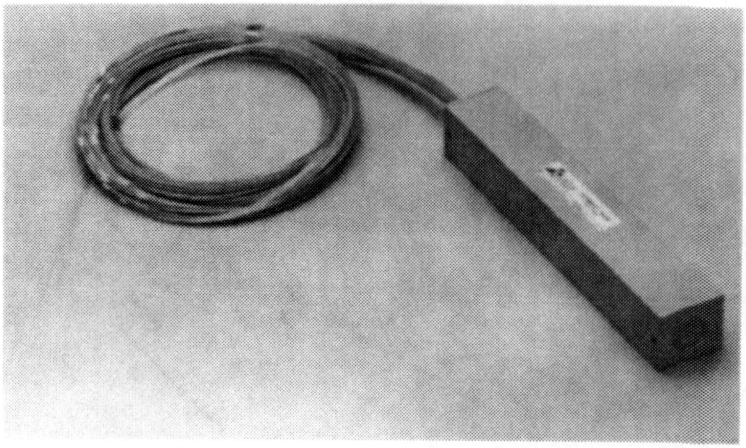

Figure 6.20 Typical diffraction grating router. 22 × 22 Input/Output ports. Spacing between channels: 1.6, 0.8, or 0.4 nm. Physical size 234 × 35 × 35. (*Source:* HighWave Optical Technology. Reprinted with permission).

Out / In	j = 1	j = 2	j = 3	j = 4
i = 1	λ_2	λ_3	λ_4	λ_5
i = 2	λ_3	λ_4	λ_5	λ_6
i = 3	λ_4	λ_5	λ_6	λ_7
i = 4	λ_5	λ_6	λ_7	λ_8

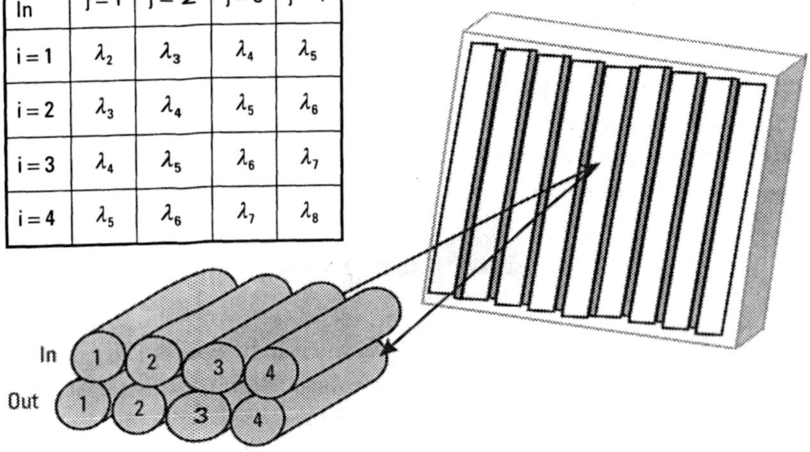

Figure 6.21 Diffraction grating router with a double fiber array at spectral focus. (Remark: When the grating is used in a single diffraction order, 2N − 1 wavelengths are necessary to connect N inputs to N outputs, here 7 wavelengths for a 4 × 4 router. Using different orders N − 1 wavelengths can be enough for nonblocking connectivity. New designs will allow N × N cyclic routing with N wavelengths only.

Out / In	j = 1	j = 2	j = 3	j = 4
i = 1	λ_2	λ_3	λ_4	λ_5
i = 2	λ_3	λ_4	λ_5	λ_6
i = 3	λ_4	λ_5	λ_6	λ_7
i = 4	λ_5	λ_6	λ_7	λ_8

Figure 6.22 Diffraction grating router with a single fiber array at spectral focus (same remark on wavelengths as in Figure 6.21).

We first assume a single array of fibers (Figures 6.22 and 6.23). A set of p fibers is located in the focal plane F of an optical system composed of a focusing lens and a grating, the fibers are numbered from 1 to p, i and j are the references of any fiber pair. If λ_{ij}, the wavelength in the medium in front of the grating, corresponds to the coupling between the entrance fiber i and the exit fiber j in the first order of the grating in a diffraction order m, considering a distance between grooves on the grating a, we obtain:

$$a(\sin\alpha_i + \sin\alpha_j) = m\lambda_{ij}$$

If the position of the fibers in the focal plane is such that: $\sin\alpha_{i+1} = \sin\alpha_i + u$, and $\sin\alpha_{j+1} = \sin\alpha_j + u$ where u is a constant, then we obtain with a grating working in a fixed diffraction order: $\lambda_{ij} \approx \lambda_0 + (i + j)\,\Delta\lambda$. It is easily demonstrated that $\lambda_{ij} = \lambda_{ji}$ and that $\lambda_{ij} = \lambda_{i+p,\, j-p}$. Thus, Table 6.3, corresponding to λ_{ij} in a given order m, is symmetrical relative to its diagonal and the wavelengths corresponding to the straight lines such as i + j = constant are equal. Note that the wavelengths such as λ_{i+j} are given by an arithmetical progression of a × u difference.

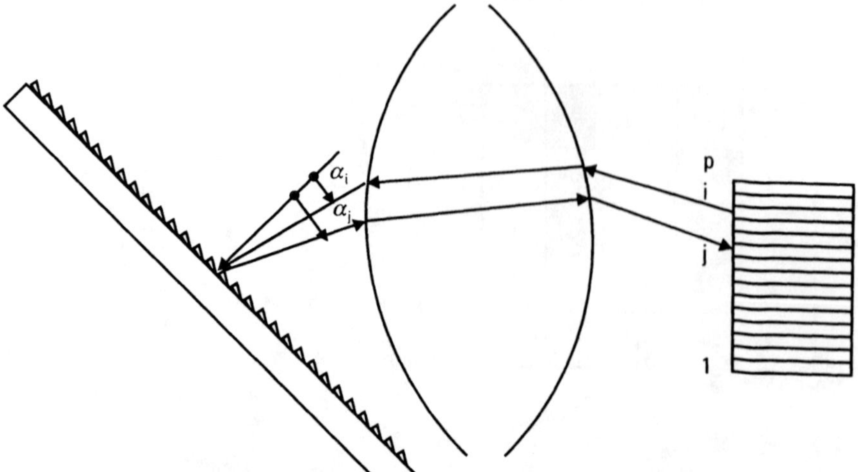

Figure 6.23 Schematic view of a grating wavelength router (multiplexer) with its focusing lens and the fiber plane. In mountings with a plane grating, the rays coming from each fiber are generally parallel in front of the grating. In mountings with a concave grating, the focusing lens is replaced by the grating itself. This is also the case in AWG in which the grating is replaced by the waveguide array.

Using two different diffraction orders, a cyclic permutation of wavelengths can be obtained: $\lambda_{ij} \approx \lambda_0 + (i + j)\ \Delta\lambda$ when $i + j \leq N + 1$ and $\lambda_{ij} \approx \lambda_0 + (i + j - N)\ \Delta\lambda$ when $i + j \geq N + 2$. So instead of $2N - 1$ wavelengths for a $N \times N$ router N wavelengths would be enough. However, this needs to work in high orders (typically 59 and 60 in C band).

Table 6.3 gives entrance to exit fiber wavelength relation in a static router with constant spacing between fibers in the focal plane. Note the symmetry relative to the diagonal $i = j = 1$ to $i = j = p$ and the fact that the wavelength is unchanged along straight lines perpendicular to the diagonal. λ_{ij} linearly increases with $i + j$.

A Typical Plane Grating Device

In [127, 128] 22 × 22 and 25 × 26 static routers using diffraction gratings with 0.8-nm channel spacing were demonstrated. They were interleaved, in order to get 47 × 48 routers working on a 0.4-nm wavelength grid. In these devices the double fiber array is arranged in silicon V grooves. The entire device is constructed in a solid silica block (Figure 6.24). A plane diffraction grating inserted inside the silica block positions the wavelengths coming from any input fiber of the first row of fibers to any output fiber of the second row of fibers (as in Figure 6.20).

A double array of 25*26 SM fibers with 42-mm spacing on both sides of silicon chips is shown in Figure 6.25. A fiber-positioning accuracy better than 1 μm was demonstrated. The contribution of V groove error of position was less than 0.5 mm. The main error came from the core-to-cladding offset tolerance of the fibers. This gave measured wavelength accuracy for each

Table 6.3
Wavelength Used to Connect Any Input Fiber to Any Output Fiber

	Fiber Out 1	**Fiber Out 2**	**Fiber Out 3**	—	**Fiber Out j**	—	**Fiber Out p**
FIBER IN 1	λ_{11}	λ_{12}	λ_{13}				
FIBER IN 2	λ_{12}	λ_{13}	λ_{14}				
FIBER IN 3	λ_{13}	λ_{14}	λ_{15}				
—							
FIBER IN i					λ_{ij}		
—							
FIBER IN p							λ_{pp}

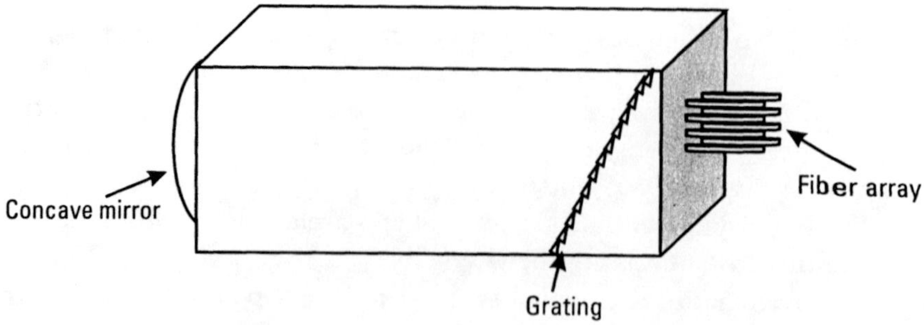

Figure 6.24 Diffraction grating router configuration. A plane diffraction grating inserted inside the silica block positions the wavelengths coming from any input fiber of the first row of fibers to any output fiber of the second row of fibers. Here, a first fiber row (i = 1 to 4) is imaged through a grating to a second fiber row (j = 1 to 4): $\lambda_{ij} \approx \lambda_0 + (i + j)\Delta\lambda$.

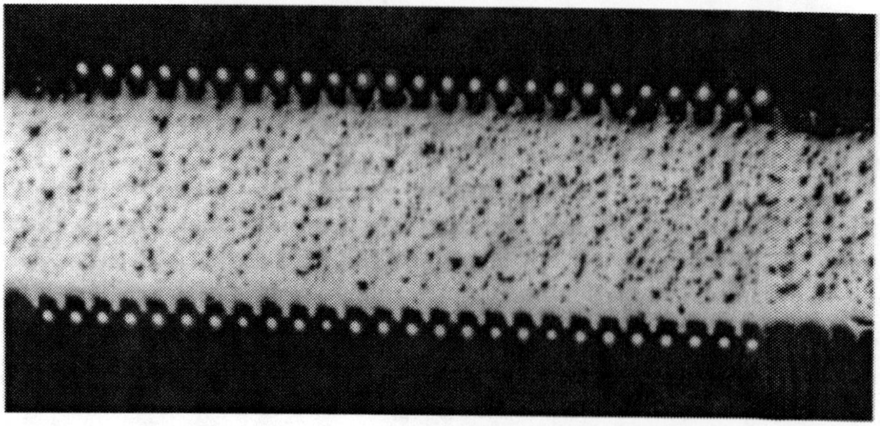

Figure 6.25 Microscope view of a double array of 25*26 SM fibers in silicon V grooves, with 42-mm spacing. (*Source:* Highwave Optical Technologies.)

channel of 0.025 nm for 0.4-nm spacing. The housing dimensions of the device was 234*35*35 mm. The FWHM of transmission channels was 0.22 nm, and the insertion losses was 2.4 dB, with a crosstalk of −45 dB. All 650 bandpasses from any 25 inputs to any 26 outputs had low losses and almost identical Gaussian shapes as calculated [119]. Lateral aberrations in the field were less than 0.3 μm. It was demonstrated that routers with wavelength spacing as small as 0.05 nm and with more than 100 × 100 ports could be made in this configuration.

A Typical Concave Grating Device

Impressive results were published in [130]. A 91 × 91, 0.33-nm spacing, free-space WDM router based on a new concave grating configuration giving a loss of less than 1 dB and crosstalk level lower than −52 dB was reported (Figure 6.26).

All these results show that free-space or monoblock diffraction-grating technology with simple or double arrays of fibers can be used for manufacturing very dense static N*M wavelength routers, with very small channel spacing. Down 0.05-nm spacing with wavelength interleaving and perhaps less would be feasible but other problems such as nonlinear crosstalk, sources and other device stability, wavelength line-width enlargement with modulation, and so on, limit the interest in such ultra-dense routers.

6.7.2.4 Static Wavelength Router Based on AWG Configuration

Using N input ports and M output ports with an AWG [132, 133] an N × M static wavelength router can be obtained [134, 135].

With an input fiber i corresponding to an angle Φ_i in the input slab waveguide and an output fiber j corresponding to an angle θ_j in the output slab waveguide (see Figure 6.27), the wavelength selected in the output fiber is $\lambda_{ij} = 1/m[n_s d\,(\sin\Phi_i + \sin\theta_j) + n_a\,\Delta L]$ where m is the diffraction order, n_a is the effective index of the arrayed waveguides, n_s is the effective index of the slab waveguides, d is the distance between two adjacent waveguides in the array, and ΔL is the length difference between adjacent guides.

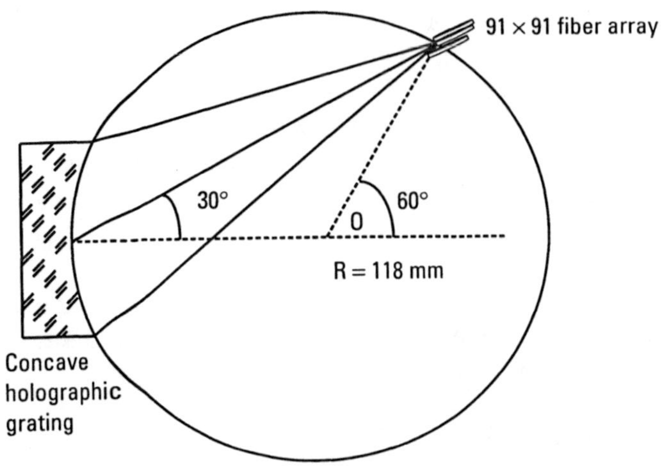

Figure 6.26 Free-space concave holographic grating router configuration. (*After:* [130].)

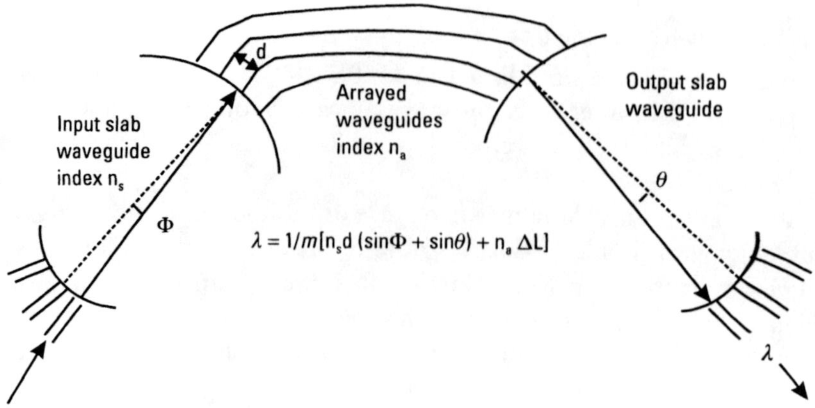

Figure 6.27 Static wavelength router based on AWG configuration.

As for diffraction-grating devices, if the position of the fibers in the focal plane is such that $\sin \Phi_{i+1} = \sin \Phi_i + u$, and $\sin \theta_{j+1} = \sin \theta_j + u$ where u is a constant, then we obtain with a grating, working in a fixed diffraction order: $\lambda_{ij} \approx \lambda_0 + (i + j)\,\Delta\lambda$. As with diffraction gratings we get $\lambda_{ij} = \lambda_{ji}$ and $\lambda_{ij} = \lambda_{i+p,\,j-p}$. Thus, the table corresponding to λ_{ij} in a given order m is symmetrical relative to its diagonal and the wavelengths corresponding to the straight lines such as $i + j =$ constant are equal.

Using two different diffraction orders a cyclic permutation of wavelengths can be obtained: $\lambda_{ij} \approx \lambda_0 + (i + j)\,\Delta\lambda$ when $i + j \leq N + 1$ and $\lambda_{ij} \approx \lambda_0 + (i + j - N)\,\Delta\lambda$ when $i + j \geq N + 2$. With the high diffraction orders used in AWG it is easy to get these conditions without additional losses [136].

But it is not possible to obtain a perfect cyclic wavelength or frequency response because the free-spectral range varies between the different orders. This problem can be reduced with specific adjustments of fiber spacing in the input and output fiber arrays. For example in [137] a cyclic 16×32 AWG router at 100-GHz spacing with a center wavelength deviation reduced down to ±0.05 nm is reported. This is obtained with two sets of fiber spacing in the input and in the output fiber arrays: in this component, in the input array 30.74 and 31.02 μm from input 1 to 11 and from input 11 to 16, respectively, and in the output array 30.60 and 30.85 μm from output 1 to 18 and from output 18 to 32, respectively.

6.7.2.5 Interleaving of Passive Wavelength Routers

As with elementary wavelength multiplexer/demultiplexers, it is possible to associate static wavelength routers working on interleaved wavelength grids.

With two N × N routers working with the same Δf frequency spacing but with an Δf/2 shift between the two grids, it becomes possible to design 2 N × 2 N–routing devices working on an Δf/2 frequency grid [Figure 6.28(a)]. This architecture is useful but not fully nonblocking. With four N × N routers in a switch-expansion architecture (see Figure 6.8), 2 N × 2 N–fully nonblocking can be designed. Two N × N routers working one in each direction can be made with the two N × N interleaved routers [Figure 6.28(b)]. This architecture is also fully nonblocking.

6.7.2.6 Static Waveband Routers and Application to Waveband Cross-Connection

In high-transmission-capacity large telecommunication networks, petabit/s-aggregate-capacity cross-connection between many fibers (100 and more) with many wavelengths (100 and more) per fiber, each at high bitrates (10 Gbps and more), will have to be done. 1,000 × 1,000 to 10,000 × 10,000 switches may be necessary if the switching is to be done at the individual wavelength-fiber level. In the "multigranularity" approach, grouping as many wavelengths which have a common destination as possible,

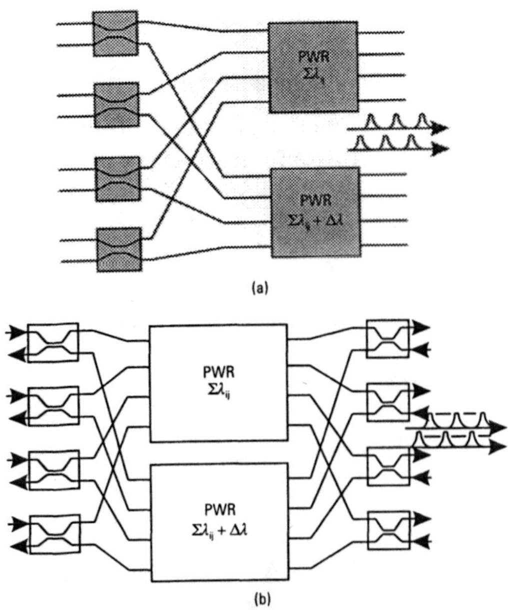

Figure 6.28 Interleaving of routers. In (a), an 8 × 8 router is made with two 4 × 4 interleaved routers. In (b), two 4 × 4 routers, one working in each direction, are made with the same two 4 × 4 interleaved routers; (b) is fully nonblocking.

allows reduction of the number of ports of the cross-connector switches [138–141]. This WDM grooming is to some extent equivalent to the TDM grooming: The optical granularity can be considered as an extension of the electrical granularity. The routing and wavelength assignment process will try to avoid unnecessary demultiplexing of the fibers in their individual bands of wavelengths, and the bands of wavelengths in their individual wavelengths at the nodes.

A typical optical cross-connector in big nodes will be made of an association of a fiber cross-connector, a waveband cross-connector (with a different set of wavelengths in each waveband), and a wavelength cross-connector.

A diffraction grating or an AWG static router with input fiber spacing larger than output fiber spacing, corresponding to wavelength spacing $\Delta\lambda$ and $d\lambda = \Delta\lambda/m$, respectively, can be used to address predetermined sets of wavelengths from any input to any output as shown in Table 6.4 and Figure 6.29.

This component can also be used as a demultiplexer/multiplexer of bundles of wavelengths as shown on Figure 6.29. Such components can be used as static routers in passive optical-network architectures with switching functionality relegated to the periphery or associated with active space switches in the design of waveband cross-connectors.

6.7.2.7 Problems in Getting All Channels on the ITU Grid

We have seen above that wavelength λ_{ij} from input i to output j is given by a similar law in a diffraction grating and in an AWG device:

$$\lambda_{ij} = (1/m)\left[a\left(\sin\alpha_i + \sin\alpha_j\right)\right] \text{ for a diffraction-grating WDM.}$$

$$\lambda_{ij} = (1/m)\left[n_s d\left(\sin\Phi_i + \sin\theta_j\right) + n_a \Delta L\right] \text{ for an AWG WDM.}$$

Table 6.4
Static "Waveband" Router

In	Out A	B	C
A		$\lambda + \Delta\lambda/3$	$\lambda + 2\Delta\lambda/3$
B	$\lambda + \Delta\lambda$	$\lambda + \Delta\lambda + \Delta\lambda/3$	$\lambda + \Delta\lambda + 2\Delta\lambda/3$
C	$\lambda + 2\Delta\lambda$	$\lambda + 2\Delta\lambda + \Delta\lambda/3$	$\lambda + 2\Delta\lambda + 2\,\Delta\lambda/3$
D	$\lambda + 3\Delta\lambda$	$\lambda + 3\Delta\lambda + \Delta\lambda/3$	$\lambda + 3\Delta\lambda + 2\Delta\lambda/3$

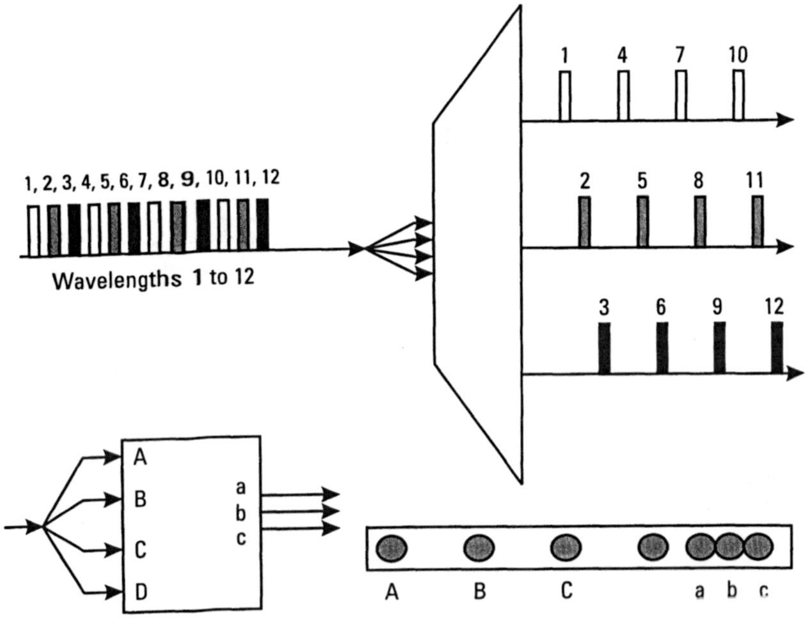

Figure 6.29 Static "waveband" router used to demultiplex groups of wavelengths. A 1/N divider or a wavelength interleaver (lower losses) is located between the input and ABCD.

Then, in the approximation of small angles for equidistant fibers, we have $\lambda_{ij} \approx \lambda_0 + (i + j)\ \Delta\lambda$ in a given order on both devices. In the table of wavelengths λ_{ij} we get channels equally spaced in wavelength, in a given order, in this approximation. However, the ITU channels are defined as equally spaced in frequency (f). For this, it is possible to design fiber arrays with nonequidistant fibers. For instance, we can position the fibers in order to get f_{ij} on the ITU grid for all i = j, but for all other f_{ij} we get a mismatch from the ITU grid. The mismatch is maximum on f_{1N} and f_{N1}. Other arrangements of fibers can be used, but it is not possible to get all channels of the router perfectly on the ITU grid. This is why, for large number of ports, the N optical channels (typically 80 × 80, 0.4-nm spacing) should be chosen equally spaced in wavelength and not in frequency [142].

We have also in any case $m'\lambda_{ijm} = m\lambda_{ijm}$ for wavelengths from the same input i to the same output j in two different orders m and m'. So, for a given couple of input to output fibers, we get the frequencies f_m on a grid with Δf constant. In other words, the free-frequency range Δf is proportional to the diffraction order but the free-spectral range $\Delta\lambda$ is in inverse proportion of the diffraction order. So it is not possible to design large cyclic-grating routers, using two diffraction orders, with the same wavelength spacing in the two different orders, nor static grating routers (cyclic or not) equally spaced in frequency for any (i,j) couple. In a given order it is possible to get a constant spacing in wavelength with a very good approximation. This approximation is sin $\propto \approx \propto$, for small $\propto$ angles. As it is possible to keep the same angular dispersion, with smaller angles and a larger focal length, the spacing in wavelength can be made as constant as necessary for all (i, j) couples. Static grating routers, equally spaced in frequency in a given order on the diagonal (i = j), are feasible. Some mismatch with the ITU grid is unavoidable for the other (i, j) couples, with a maximum error on (1, N) and (N,1) couples of input to output. Another solution is an equal frequency spacing on the ITU grid for all outputs corresponding to a given input, and for all inputs corresponding to a given output. Again, a mismatch with the ITU grid is unavoidable for the other (i, j) couples.

6.8 Optical Cross-Connector

The key role of the optical cross-connector (OXC) is to reconfigure the network at the fiber and wavelength level, for restoration or to accommodate change in traffic demand. The OXC plays a role in the optical domain similar to the role of the digital cross-connect systems (DCS) in the electronic domain. The OXC is utilized to interconnect rings or different nodes in a mesh network. There is now wide agreement that the big revolution in optical networking will occur in the form of optical-switching equipment [143].

Today, the hybrid OXC first converts the optical data to electronic data and uses DCS for the practical cross-connection. There is no other solution for very high numbers of input to output ports N × N cross-connection. N may be on the order of thousands in a DCS, but only on the order of tens in all-optical switches today. But N is already on the order of a few hundreds in optical switches at the laboratory stage.

The OXC is based on a space switch, or on a wavelength space switch: WSXC or WIXC.

6.8.1 WSXC

6.8.1.1 Principle

Sometimes they are also called wavelength selective switches. One of the wavelengths arriving on one of the N ports is demultiplexed, and goes to the

N × N switch working on this specific wavelength. This wavelength can be switched to any chosen output port after multiplexing. N demultiplexers, J switches, and N multiplexers are used in an N × N node with up to J wavelengths on each fiber (Figure 6.30).

6.8.1.2 Enabling Technology

This component can be made of individual optical multiplexers, demultiplexers, and switches. WSXC can be designed with different kinds of WDM components. It may use a multimultiplexer that replaces all multiplexers and demultiplexers in a single smaller and cheaper component using three-dimensional optics (still necessary for large N and or J) or integrated optics. For small N × J values, integrated optics components including the switches with some limitations are already available. For high values of N and J, the "free-space" optics are generally the preferred solution. Integrated WSXC can be manufactured with AWG and optical switches [144, 145]. In [146], a modular WSXC with a 256 × 256 crossbar switch consisting of 16 integrated broadcast modules and 16 integrated switch modules is proposed. Each broadcast module uses an AWG 16-wavelength multiplexer and a splitter. Each switch module uses 1 × 16 splitters, 256 SOA gates, static AWG routers, and combiners.

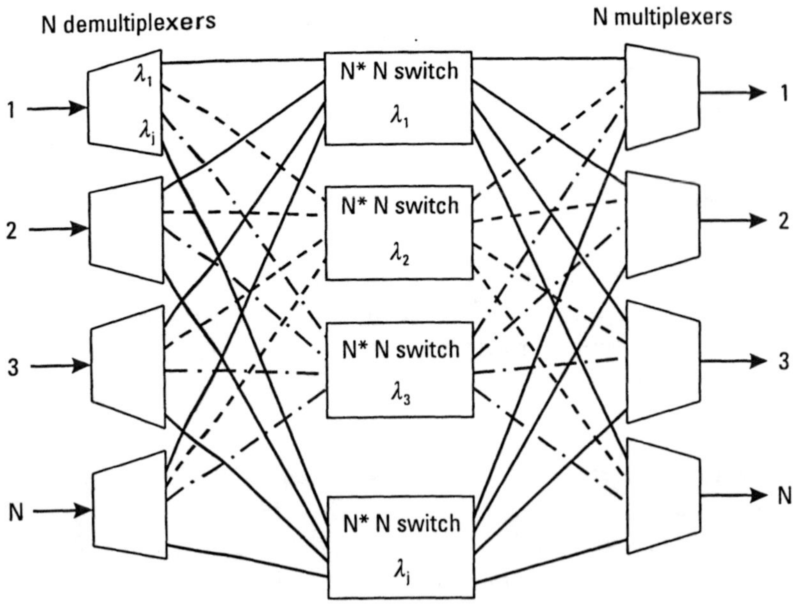

Figure 6.30 A wavelength selective cross-connector.

In [147], a 2 × 2 port, 16 wavelength 0.8-nm spacing integrated WSXC is reported. The insertion loss is 8 to 10 dB and the crosstalk is −28 dB. Details on this device are given in [148]. Two couples of AWGs and MZ-type thermo-optic switches are monolithically integrated on a common silicon substrate. This component was used as an ADM.

A very compact (6 × 6 mm) polarization-independent integrated four-channel 2 × 2-optical cross-connect on InP is reported in [149, 150]. It works on four frequencies spaced at 400 GHz. The device uses two 8 × 8-polarization-dispersion-compensated AWG, and four 2 × 2 electro-optical switches. The total loss is below 13 and 16 dB for TE and TM polarization, respectively. The crosstalk, determined by the performance of the switches, is lower than −16 dB.

In particular architectures, the switches can be replaced by active wavelength routers. For example, in [151] an active electro-optic section over an arrayed waveguide enables the AWG to become a space-wavelength switch.

Much information on integrated WSXC and other advanced devices for WDM applications is given in [152]. In particular, results with 4 × 4 four-wavelength compact devices are presented:

- 1.5 × 3-mm^2 chip with four 1 × 4 AWG and four push-pull MZI's: total on-chip loss (TE) < 12 dB, interchannel crosstalk < −18 dB.
- 9 × 12-mm^2 chip with a single 4 × 16 phasar (static router used as multimultiplexer) replacing the four AWGs and MZIs, total on-chip loss (TE) 12 to 16 dB, interchannel crosstalk < −13 dB.

It can be seen that the crosstalk is larger in the second configuration.

- 8 × 12-mm^2 chip with dilated switches, made of two 2 × 8 AWGs, a 4-wavelength 2 × 2 WSXCs, showing total on-chip loss < 16 dB, and interchannel crosstalk < −20 dB. This illustrates the interest of dilated switches to reduce the crosstalk.

WSXC with few ports and few wavelengths based on FBGs and circulators were proposed. S. -K. Liaw, S. Chi, and K. -P Ho [153] propose the four configurations of Figure 6.31 corresponding to 2 × 2 cross-connectors working in "transmission" (a), or in reflection (b), unidirectionaly, or working in "transmission" (c), or in reflection (d), bidirectionally. Strain-tunable FBGs are used for wavelength selection.

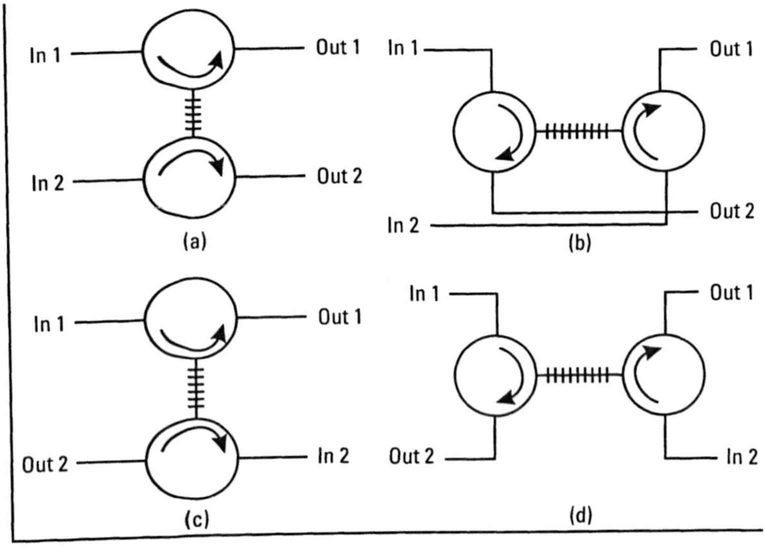

Figure 6.31 Four configurations of WSXC using FBGs: (a) unidirectional "transmission" type, (b) unidirectional "reflection" type, (c) bidirectional "transmission" type, and (d) bidirectional "reflection" type. (*After:* [153].)

Note that in specific fiber-wireless networks, T. Koonen, et al. demonstrated the use of a flexible wavelength routing using polarization-independent reflective modulators at the ONUs [154]. The system uses 8-wavelengths HDWDMs, power splitters, and thermo-optics MZ 1 × 8 switches in glass-integrated optics. The polarization dependency is less than 1 dB and the extinction ratio better than 20 dB.

Also note that a specific wavelength selective cross-connect using tunable filters and equalizers in front of space-switches between splitters and combiners is shown in Figure 6.32. In such devices with large splitting/combining losses, it is in general necessary to use optical amplifiers at each input and output.

6.8.1.3 Wavelength-Dilated Switches

Wavelength-dilated switches [155, 156] can relax the crosstalk requirement of wavelength-routing devices. In these devices, the optical switch matrix is divided into several subswitch matrices working on subsets of wavelengths with spacings that are multiples of those in the whole channel set. For example, in Figure 6.33, with a set of eight wavelengths ($\lambda_1, \lambda_2, \lambda_3, \lambda_4, \lambda_5, \lambda_6, \lambda_7, \lambda_8$) in an 8-input ports to 8-output ports node, three elementary 8 × 8 switches

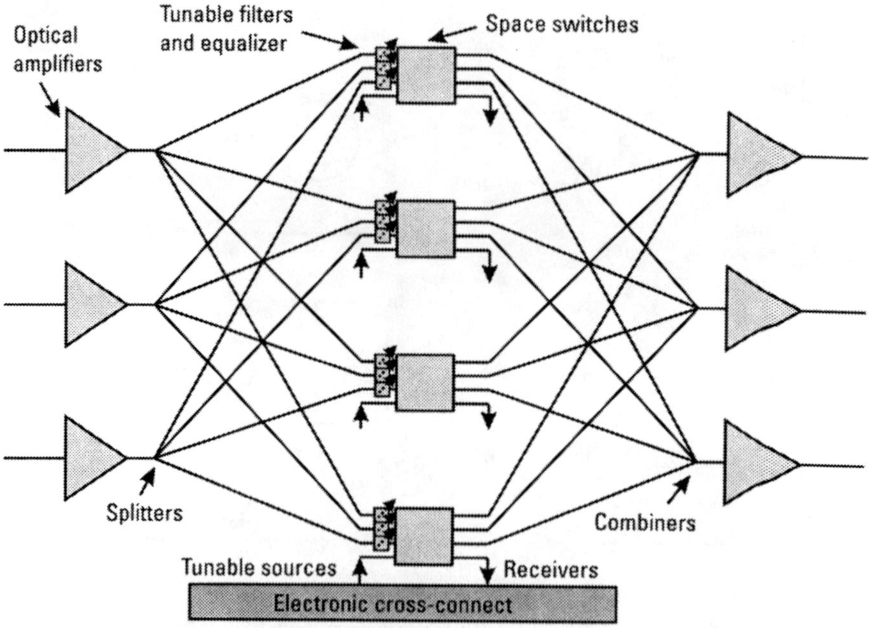

Figure 6.32 WSXC using tunable filters.

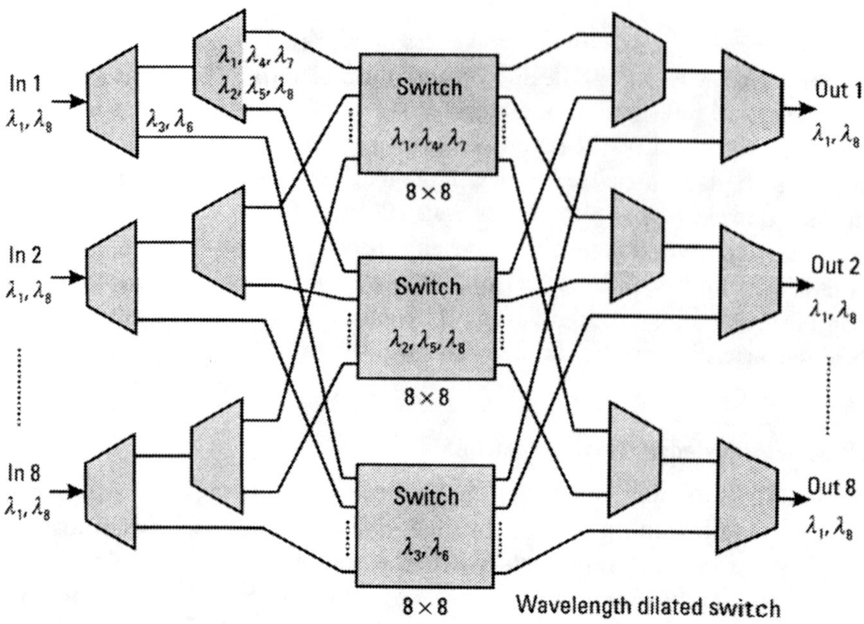

Figure 6.33 Wavelength-dilated switch.

working on ($\lambda_1, \lambda_4, \lambda_7$), ($\lambda_2, \lambda_5, \lambda_8$), and ($\lambda_3, \lambda_6$), respectively, are used to design the wavelength space switch. Then each subswitch matrix operates on a reduced number of wavelengths with larger wavelength spacing, thus reducing interchannel crosstalks.

6.8.1.4 WSXCs in a Series

In practical networks, the WSXCs are generally used in series. The polarization defects and crosstalk are generally added. However, with adequate components the penalty can be relatively low even at high bit rates. For instance, in [157] a 10-Gbps transmission through four WSXCs spaced by 70 km each on a link of 200 km of TrueWave fiber showed a measured power penalty of 1.8 dB only.

6.8.1.5 Reliability of WSXCs

In WSXCs, it is important to detect and repair any routing failure at the earliest stage. Some Bellcore or ITU documents specify the reliability-assurance requirements of given networks.

In 1999:

- GR-418 "Generic Reliability Assurance Requirements for Fiber Optics Transport Systems";
- GR-2979 "Common Generic Requirements for OADMs and OTMs";
- G.873 "Working List Contribution T1X1.5/99-111."

R. Holmstrom and L. Wosinska [158] evaluated the failure rates of some components used in cross-connects in FITs (failures per 10^9 hours): 1:2 splitter 50, tunable transmitter 745, tunable receiver 470, WDM 8 channels 360, transmitter 186, receiver 70, 16 × 16 optical matrix switch 1,000, 16 × 16 cross-point switch 320, thermoelectric cooler 1500. Of course, these values were given for typical components commercially available in 1999 and progress was expected. This gave a mean time between failure (MTBF) of about 12 years, and 6 years on 128 × 128 WSXCs without or with OEO conversion, respectively. So redundant equipment or cross-connectors with redundant paths (less expensive) were needed. They showed how the MTBF could be ameliorated with a shared redundancy: The new MTBF of the 128 × 128 WSXC was about 66 years with or without OEO conversion.

It is also very important to detect and repair any routing failure locally. For these devices, supervisory wavelength channels were proposed. In [159] or [160] the supervision is based on demultiplexing without additional losses of channels in a diffraction order $m + 1$, m being the diffraction order used for the transmission channels. In [161], one wavelength channel is reserved for transmission of the optical supervision channel and protection-switching actions in case of failure are initiated locally.

From the management system point of view, informative concepts and design for management of connection services with fast-path provisioning and restoration in the event of a failure in dynamic optical networks using cross-connects was given in [162].

6.8.2 Cross-Connect with Wavelength Conversion

6.8.2.1 Principle

The condition of wavelength continuity along the optical path, necessary in all-optical networks, may lead to a blocking when two signals with the same wavelength need to take the same fiber route. Let us take the example of Figure 6.34. It is necessary to change the original wavelength of the signal going from A to F: λ_1 needs to be converted to λ_2 in the WXC in order to avoid a conflict along their common path with the signal from C to E also at λ_1. The wavelength assignment policies and routing strategy in all-optical networks with limited-range wavelength conversion will not be discussed here. See, for instance, [163]. We will only describe some cross-connect architectures allowing full or limited wavelength conversion.

6.8.2.2 Enabling Technology

In these devices, very often the optical wavelength conversion uses MZI devices. In addition, the conversion can help reduce source chirp and ameliorate the extinction ratio.

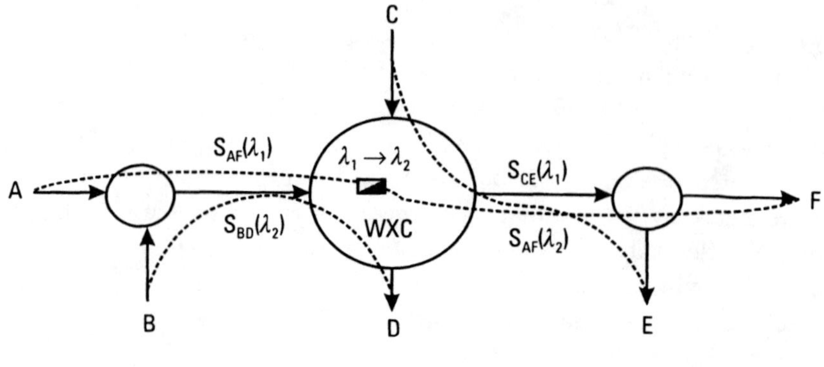

Figure 6.34 Simplified example of wavelength cross-connect with wavelength conversion.

In [164], it is emphasized that the optical wavelength conversion using InP MZI such as those reported in [165] reduced the input signal chirp and increased the extinction ratio. A chip of 0.16 nm from 2.5-Gbps DFB-laser input was reduced below the detection limit (< 0.08 nm) by wavelength conversion, and the extinction ratio was increased from 8 dB at the input to 11 dB after wavelength conversion.

Other solutions for wavelength conversion exist, such as conversion in a DBR laser. A typical example of conversion with a DBR laser is given in [166] in which four-channel 1,546.6 to 1,552.6 1-nm spacing are obtained from a signal input at 1,538 nm and routed through a phasar.

The suitability of programmable widely tunable laser transmitters for application in reconfigurable networks was tested in [167]. Tunable transmitters with an absolute frequency accuracy better than ±5 GHz reproducible over several months of testing were obtained from three manufacturers [168–170]. More than 100 channels, 50-GHz spacing were available with a tuning range of 30 nm in the wavelength range 1,530 to 1,560 nm. The power of each source was −3 dBm with a stability of ±0.5 dB (short and long term).

6.8.2.3 Architectures of Cross-Connects with Wavelength Conversion

Architectures with a Large Switch

The concept of Figure 6.34 can be generalized. With several fibers and several wavelengths used at a typical node, a single large switch can replace the different wavelength switches of the WSXC with a wavelength converter on all exit ports as shown on Figure 6.35. This is the more conventional wavelength interchangeable cross-connect architecture. (Transmitter and receiver ports connected to a local digital cross-connect are also generally added.) With N inputs, N outputs with J wavelengths each, the dimension of the switch is NJ × JN. So the switch in this architecture needs to be very large [164, 171].

However, very large switches would be unpractical. Different solutions were proposed to reduce the complexity of this solution [172–174].

Modular Architectures

Other architectures avoiding large dimension cross-connects have been proposed, for example, in the structure of Figure 6.36. Here, the light arriving at each input port is amplified and split between small space switches (4 × 4 in our example). Tunable filters (with optional equalizers) accomplish the channel selection. At the exit ports of the space switches, wavelength converters

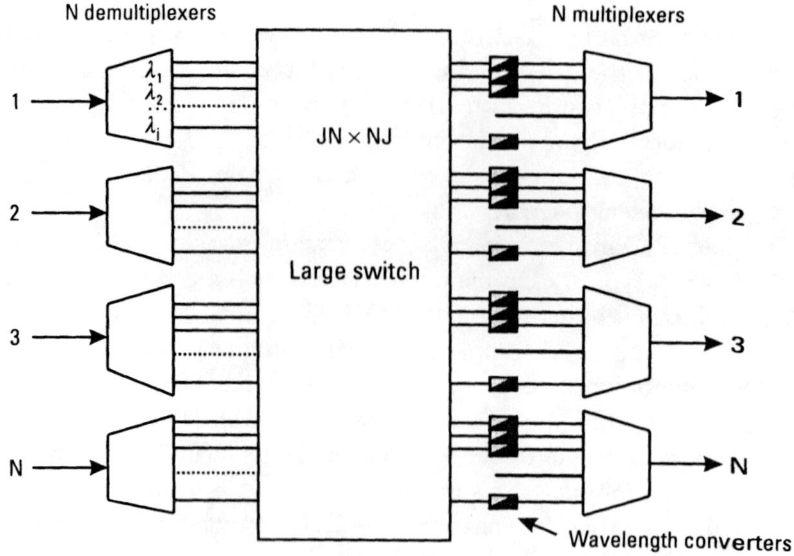

Figure 6.35 WIXC using a large switch matrix (dimension JN × NJ) and wavelength converters (JN small rectangle symbols).

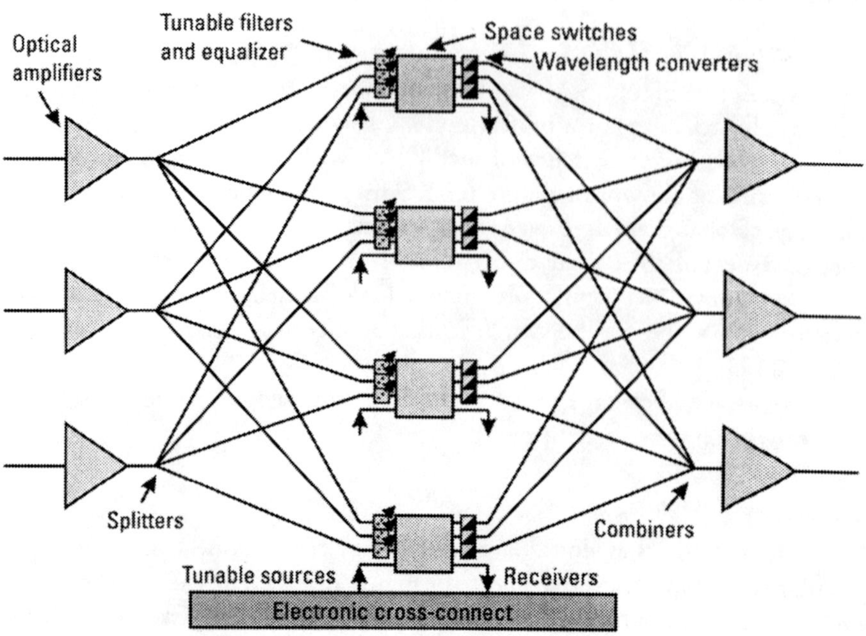

Figure 6.36 Modular WIXC with splitter and combiners.

are used and the different paths are recombined and amplified on the different output fibers connected to the node. Of course, in general, signals need to be added and dropped locally through an electronic cross-connect.

Another example is given in Figure 6.37, in which the J-incoming wavelengths in each N-input fibers are demultiplexed with one demultiplexer 1 x J per fiber, switched by a space switch J × N (with one space switch per input fiber) and combined through N, N to 1 combiners to the N-output fibers.

Such an architecture was used in [175], demonstrating a 320-Gbps aggregate capacity cross-connect system with N = 16 input and output fibers, 8 wavelengths spaced at 100 GHz, and 2.5 Gbps per channel. 1 × 8 AWG demultiplexers, 8 × 16 delivery and coupling switches, and 16 × 1 combiners are used. An O/E/O conversion with DBR lasers modulated by external modulators is used between the demultiplexers and the switches. Error-free transmission in this 5-node virtual-wavelength-path network was proved with 8-channel wavelength converted signals over 160 km without repeaters and 480 km with three repeaters.

Another version of this WIXC exists (Figure 6.38), in which splitters are used at the inputs and in which a wavelength conversion after wavelength filtering is done in the last part of the device, in front of multiplexers. This

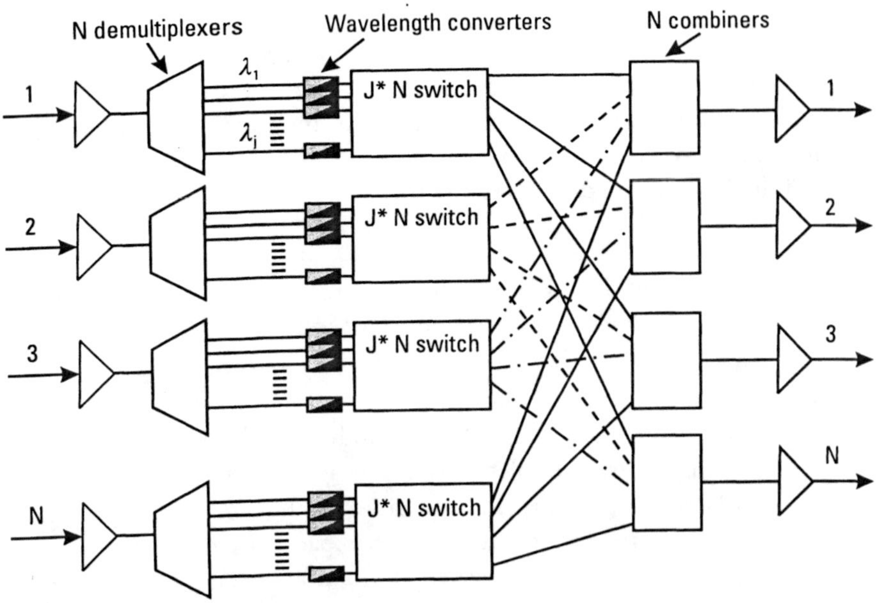

Figure 6.37 "Input-converter" WIXC with demultiplexers and combiners.

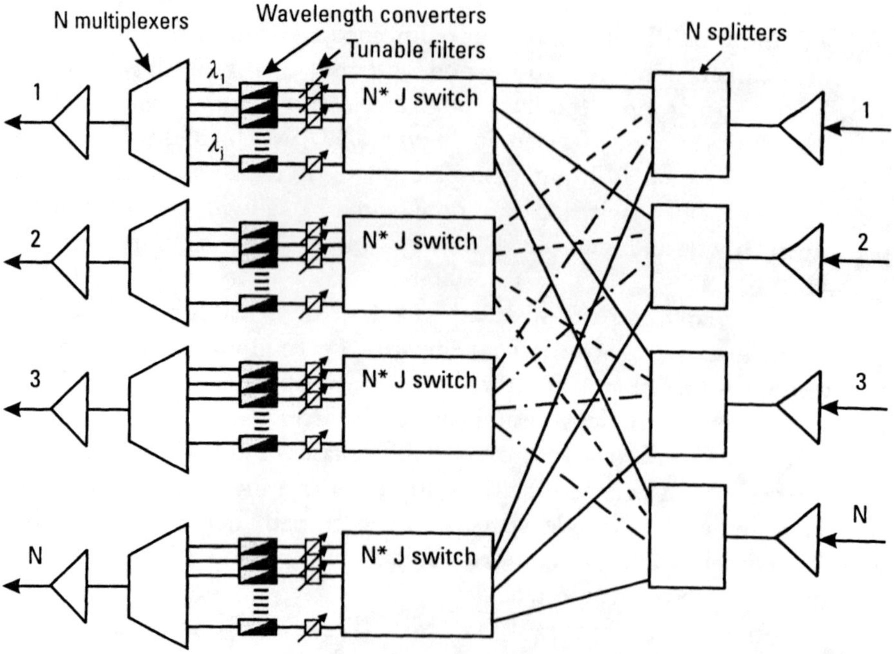

Figure 6.38 "Output-converter" WIXC with splitters and combiners.

version can be called "output-converter" architecture in opposition to the "input-converter" architecture of Figure 6.37 described above. Some advantages of the "output-converter" architectures in monitoring the switch performance are pointed out in [176].

All these structures can be modular:

- Modularity at the link level with switch, filters, and wavelength converters in each elementary module (N modules);
- Modularity at the path level with a filter and a wavelength converter in each elementary module (J × N modules). The granularity is finer but many more modules are necessary. This needs compact and affordable components. In these modules AWG or free-space routers can be used for the design of compact tunable filters that can integrate switching and wavelength filtering functions. For example, in [176], a path module for N × 1-space switching and J × 1 tunable filtering using an AWG static router with N-input ports and

(J + N – 1)-output ports, integrated with an array of 2N + J – 1 gate switches, and with a N + J – 1 combiner is proposed.

A multimodular architecture for opaque networks that is at the same time wavelength and fiber modular is presented in [177]. The possibility of introducing limited wavelength converter ability adds a flexibility degree to this design.

A practical "output-converter" cross-connect prototyped by Alcatel a few years ago is shown in Figure 6.39. It is reported in [178] that before 1997, two prototypes were manufactured. The first using proved technologies with four wavelengths each modulated at 2.5 Gbps in a 3-node self-healing network were presented in numerous exhibitions. The second with

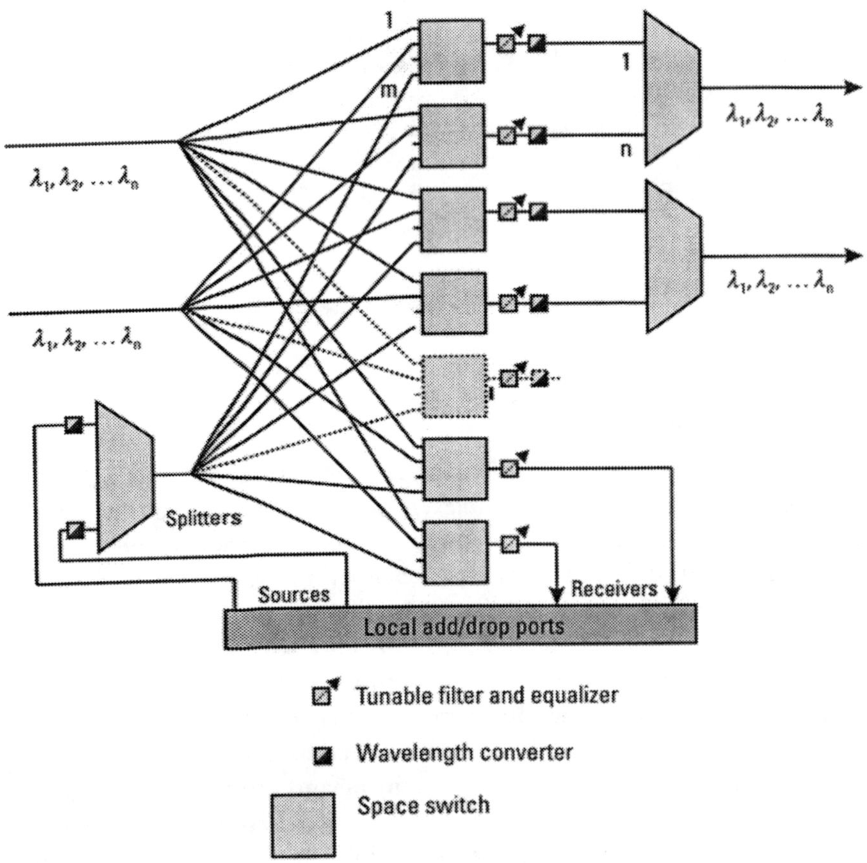

Figure 6.39 Practical "output-converter" cross-connect. (*After:* [178, 179].)

an aggregate capacity of 640 Gbps, under test in different places at that time, was a cross-connect with 4 × 4 ports, with 4-wavelengths 400-GHz spacing each at 2.5 Gbps. It used optical semiconductor gates and all optical wavelength converters (SOA-based interferometric wavelength converters) [179].

6.8.3 Layered-Switch Architecture

In fully connected WXCs, a large number of switch points could be required in the main nodes of high-capacity optical-transport networks working with tens of fibers with tens of wavelengths. However, these nodes have to connect many signals distributed in a small number of routes linked to them. It was shown that a layered-switch architecture (taking into account that a signal arriving at the node does not need full access to every signal path but only to any route) could reduce the switch size and, moreover, give a high-design flexibility and expandability. In each node the switch fabric can be separated into different layers and the layers of adjacent network nodes are cross-connected. The number of layers does not depend on the number of wavelengths as above. "The dimension of each switch can be greatly reduced to a manageable size, and the switch fabric is readily scaleable by inserting more layers" [180]. See Figure 6.40.

Crosstalk Limitations in WSXCs

Optical cross-connects generate crosstalks on the signal. If the crosstalks come from channels at the wavelength of this signal, they are generally called intraband crosstalks. If they come from other wavelengths then they are called interband crosstalks. This problem received much attention (see, for instance, [181–189]).

A typical WSXC with J wavelengths, N inputs and N outputs is shown in Figure 6.41. It consists of N demultiplexers, J N × N switches, and N demultiplexers. On a wavelength λ_{iq} coming from port i on wavelength q and connected to a given output, different crosstalks are superimposed. Crosstalks in each demultiplexer give J − 1 interband crosstalks on the signal output. For instance, on λ_{11} we get a crosstalk $\Sigma_q \lambda_{1q}$ (q from 2 to J) in the demultiplexer connected to input 1. These interband crosstalks will be double filtered in the demultiplexer and in the multiplexer in this device, but sometimes star couplers are used in place of the demultiplexers, so that there is no filtering in the second part of the system. N − 1 intraband crosstalks are generated directly in switch λ_1 and additionally J − 1 intraband crosstalks λ_{i1} on signals λ_{iq} coming from the crosstalk in the demultiplexers connected to the multiplexer used by λ_{11} are added at the output port 1. They are also generally

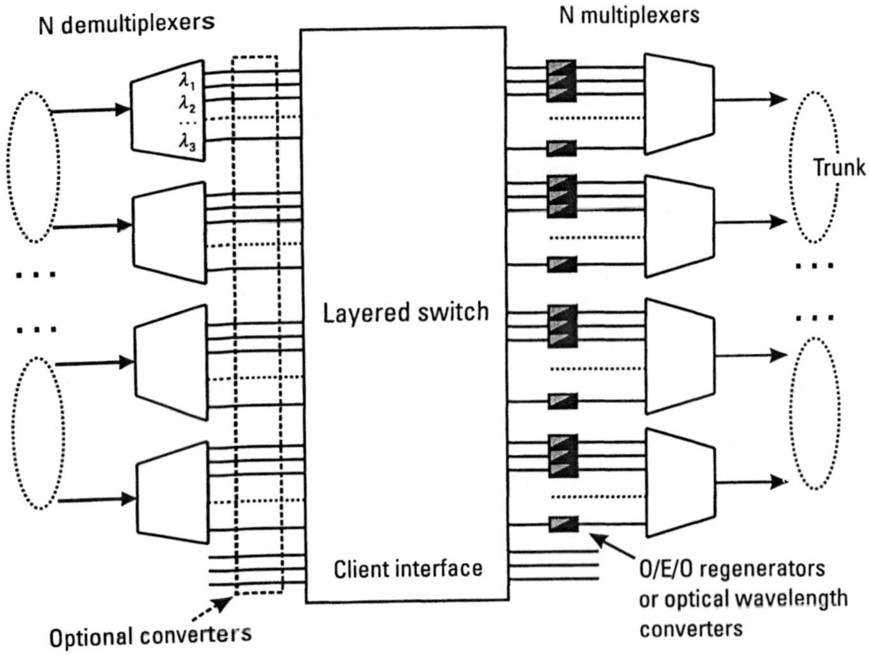

Figure 6.40 Layered-switch architecture. (*After:* [180]).

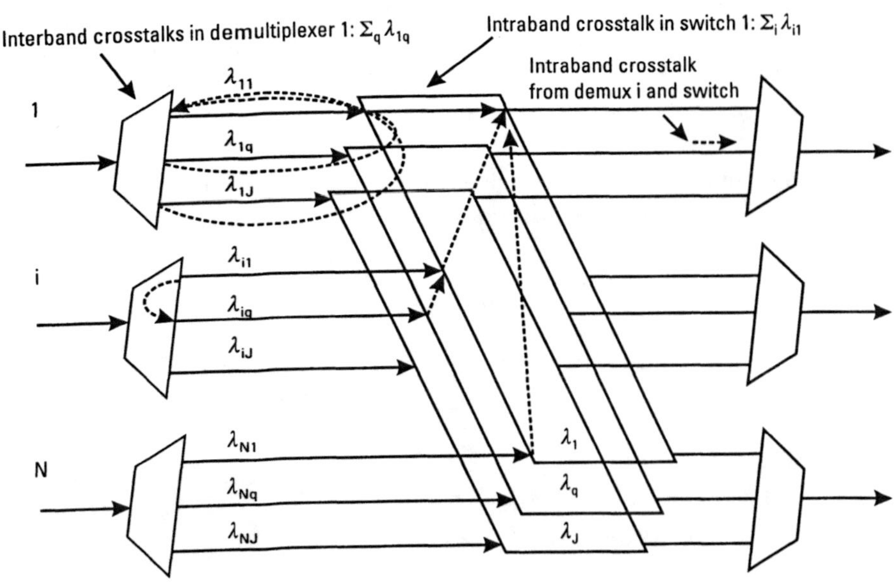

Figure 6.41 Crosstalk in WSXC.

double filtered in the demultiplexer and in the multiplexer. The typical crosstalk of some components that are used in WSXCs is given in Table 6.5.

Unfortunately, the intraband crosstalk is accumulated at each node of the network. Consequently, in networks with some nodes the total crosstalk coming from each cross-connect needs to be very small: < –40 to < –60 dB depending of the type of network, number of nodes, and overall specifications.

However, sometimes the crosstalk arising from the limited wave-band rejection in the wavelength selectors of the WSXC and WIXC can be measured and partially canceled. For example, in [190] the optical carrier at each wavelength is modulated with a subcarrier tone unique to that wavelength. Measured on their tones, the crosstalks coming from all adjacent channels can be canceled by weighting and summing the interfering photocurrents. With Gaussian passbands, an increase of 30% to 50% of channel density can be achieved at BER < 10^{-9} even if the initial crosstalk was higher than –10 dB.

6.9 OADMs

6.9.1 Introduction

The OADM (Figure 6.42) is a unit that selectively removes one wavelength λ_i from a multiplicity of wavelengths $\lambda_1, \lambda_2, \ldots \lambda_i, \ldots \lambda_N$ multiplexed on an incoming fiber, bypasses all other wavelengths, and adds the same wavelength generally with another data content on the transmission fiber.

More generally, the OADM can be defined as a component that:

- Demultiplexes some wavelengths from an incoming fiber and drops them locally with or without optoelectronic conversion;

Table 6.5
Typical Components Used in a WSXC

Component	Adjacent Channel dB	Other Channels dB
Filter 100-GHz spacing	–25	–35
AWG 100-GHz spacing	–30	–30
Free-space static router 100-GHz spacing	–45	–55
Mechanical switch	–60	–65
4x4 active InP switch (scalable)	–40	–40

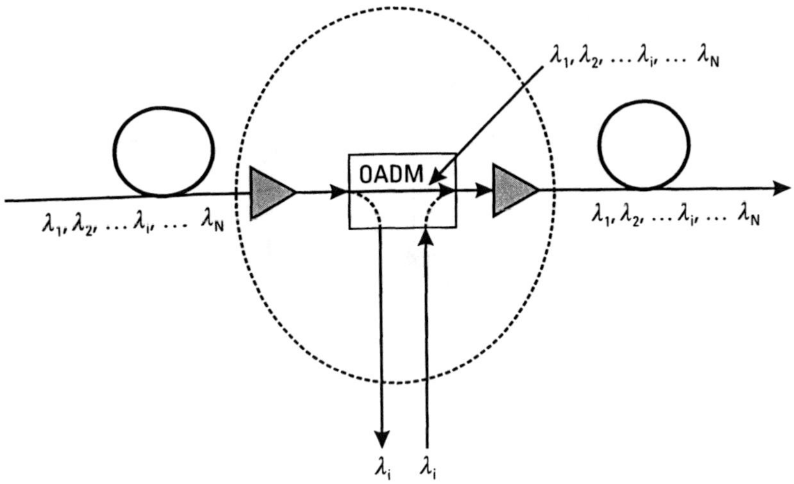

Figure 6.42 OADM principle.

- Bypasses the other wavelengths arriving from the incoming fiber to an outgoing fiber;
- Adds wavelengths from local subscribers on the outgoing fiber through a wavelength division multiplexer or a combiner;
- Demultiplexing and multiplexing of the bypassed wavelengths with the dropped wavelengths and/or with the added wavelengths is often necessary (Figure 6.43).

The OADM is either dynamically wavelength-selectable or the wavelength is fixed.

Of course, the most general OXC also includes similar functions and many of the enabling-technologies for OADM and OXC are equivalent. They will not be explained again when unnecessary.

These technologies take advantage of dielectric multilayer coatings, bulk-type or AWG devices, BGF, grating coupled waveguides, MZ, FP, and other interferometric devices.

The first OADM prototypes were designed in approximately 1980, but the early commercial OADM networks were deployed only a few years ago. Until 2000, they were mainly used on terrestrial interexchange and submarine networks. However, they now find increasing applications in metro/access networks. They are often used to bypass local exchanges.

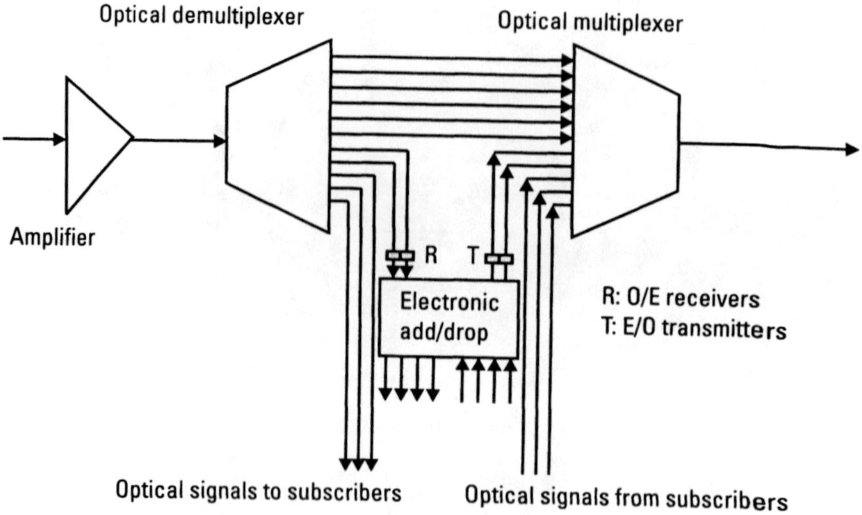

Figure 6.43 Typical OADM.

The global consumption evaluated by Electronicast was $17.22 million in 1998, with an expected growth rate of more than 100% during the next 5 years [191].

6.9.2 OADM with FBGs and Circulators

A few channel OADMs can be made with FBGs inserted between circulators as shown for a single channel in Figure 6.44. The WDM input passes through a first circulator. A wavelength channel is reflected by a Bragg grating and dropped by this first circulator. Similarly, a channel at the same wavelength can be added through the second circulator to the fiber grating which couples the signal back to the outgoing fiber.

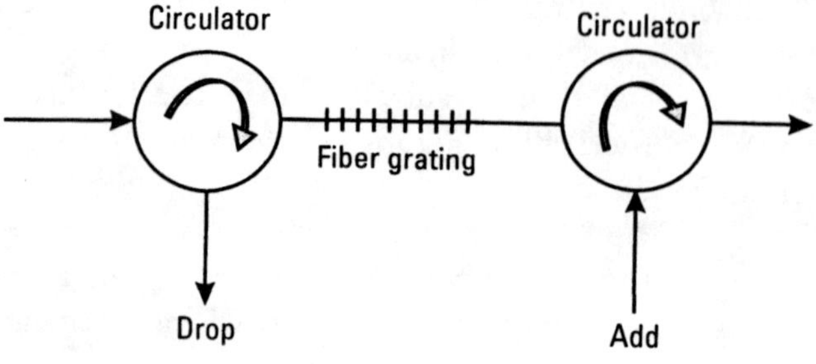

Figure 6.44 Fiber grating/circulators OADM.

Low insertion loss (2.5 dB typical) can be obtained [192] (see other references in Chapter 3). A practical application of such devices in metropolitan networks is described in [193].

The main technical problem is that the thermal drift, which is about 12 pm/°C for silica, needs to be compensated for if the temperature is not stabilized.

The channel passband can be broadened with a chirp along the grating. This chirp as well as the mean wavelength may be tunable with piezoelectric, magnetostrictive, or thermal gradient transducers. An example of the last type is given in [194], which describes an 8-channel 100-GHz-spacing tunable drop module. The chirp is achieved by two thermoelectric Peltier units set at different temperatures at the extremities of the FBG. A wavelength shift of 0.4 nm and a broadening to 0.25 nm is reported. Over the adjacent channel passband, isolation was reported as being better than 22 dB.

A technical issue of these OADMs is the crosstalk on the added channel, arising from the light remaining in the transmission path after an unperfected grating reflection R of the drop channel. The transmitted crosstalk is $T = 1 - R$, so the ratio of crosstalk to add signal is $(1 - R)/R$.

The corresponding power penalty is [195]:

$$\text{Interferometric crosstalk} \quad P = -10\log\left(1 - 4\sqrt{(T/R)}\right)$$

$$\text{Noninterferometric crosstalk} \quad P = -10\log\left(1 - 2(T/R)\right)$$

The interferometric crosstalk is much larger than the noninterferometric crosstalk. W. Bo, et al. demonstrated that a reflectivity of the grating of 99.7%, which is quite difficult to achieve, would already give a 1 dB power penalty. They discuss how to increase the reflectivity and how to reduce the crosstalk with cascaded gratings. With two gratings with a peak reflectivity of 95% each, only T = 0.25% of the drop channel goes through the gratings onto the add channel. The measured crosstalk was less than −35 dB.

6.9.3 Acousto-Optic Add/Drop

The general principle of acousto-optics used in switching devices was given in Section 6.6.3.6. In add/drop filters in general, a surface acoustic wave is applied on two transparent optical waveguides placed side by side in close proximity (Figure 6.45). A periodic modulation of the optical index is created along the waveguides by the acoustic wave. Light with wavelengths λ_1,

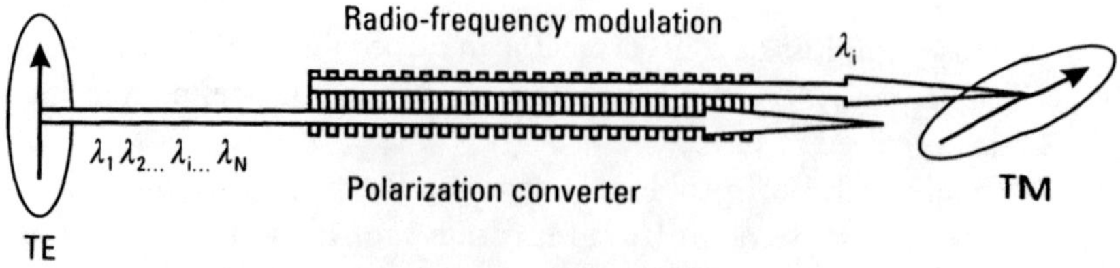

Figure 6.45 Acousto-optic OADM, simplified principle.

$\lambda_2, \ldots \lambda_i, \ldots \lambda_N$, arriving in the first guide in a given polarization (here, TE for instance), is converted in the other polarization, and coupled in the second waveguide for wavelengths that comply to the Bragg condition in the periodical index structure. Acousto-optic OADMs can be made polarization-independent in more complex designs. The devices are generally tunable.

An issue to be addressed is the quasi-coherent "intraband" crosstalk on the added signal coming from the small part of the dropped signal that can pass to the exit line directly through the component.

Different solutions were proposed to solve this problem. In [196], for instance, a special dilated double-Y structure reduces the intraband crosstalk below –30 dB on 400-GHz (and smaller) spacing OADMs. Another example is reported in [197]. It consists of a 400-GHz acousto-optic tunable filter in LiNbO3, used with a 25-GHz interleaver. Here a reduction of crosstalk below –35 dB is achieved with a double pass configuration.

6.9.4 Add/Drop with AWG

Of course, the general OADM design of Figure 6.43 can be made with two AWGs [198, 199], or a single double AWG, with a static router (see multiple multiplexers and static routers in Section 6.7), or with a single AWG in double pass.

An example of the use of static routers is given in [200]. A transmission network over 500 km with 3 nodes and an aggregate capacity of 100 Gbps uses a 32 × 32 static router in each node: Ten channels, 100-GHz-spaced, can be added or dropped in the intermediate node distant of 320 and 180 km from the two other nodes.

In a unique configuration, Y. Zhao, et al. [201] demonstrated how to eliminate the intraband crosstalk using a double-pass AWG, fiber gratings, and circulators. Through the additional filtering of the fiber gratings the accumulated amplified emission of the EDFAs is suppressed. The intraband crosstalk is –47 dB, small enough to accommodate 100 nodes (Figure 6.46).

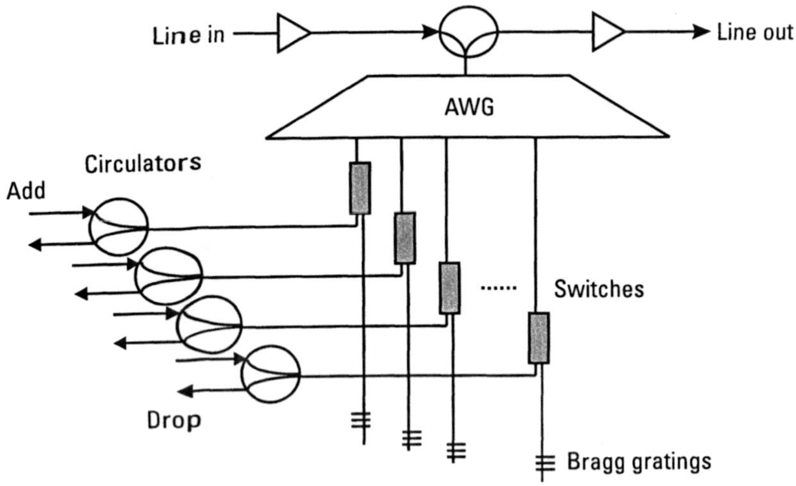

Figure 6.46 OADM with AWG in reflection with negligible intraband crosstalk. (*After:* Zhao, et al. [201].)

6.9.5 Add/Drop with Free-Space Grating Solutions

In general, the designs described for AWG are also valid for diffraction gratings including solutions with static routers and multiple WDMs. The three dimensions of diffraction-grating solutions compared to the two dimensions of AWG adds other facilities. For instance, free-space grating wavelength-selectable add/drops with tilting micromirrors were proposed. They allow a fast (ms) dynamic reconfiguration of add/drop nodes. In [202], a 16-channel OADM using a 600-lines/mm diffraction-grating device with a selectable micromirror array (MEM) in the spectral plane allows a selectable transmission of signals from an input to an output line or local add/dropping depending of the micromirrors angles. Circulators are used on the output ports to separate light coming from or going to the device: A first circulator separates the incoming and the outgoing signals, and another circulator separates the add and the drop channels. (For a review on MEMs, see [203].) Similar devices using a holographic switch [204] would be feasible and potentially faster.

6.9.6 Tunability of OADM

An important factor is the tunability of the devices. In [205], the performance of an integrated optical tunable add/drop filter (tuning over few tens of nanometers) in LPCVD SiO2 technology was reported. The losses are small enough (2 dB), however the crosstalk isolation of passing channels, −20 dB, needed further improvements.

MZI cascaded in silicon oxynitride (SiON) waveguides in a 75-mm-long device [206] give a true wavelength tunability (free-spectral range 19.2 nm) based on thermo-optic effect, and low intrachannel crosstalk (−55 dB).

The integrated OADM using grating-coupled semiconductor waveguides described in [207] demonstrates stable characteristics over 11-nm tuning range. The losses are quite high (9 dB) but very stable over tuning.

6.9.7 Cascading of OADM

In real networks such as SONET ring networks, these components have to be cascaded, generally one in each node. Problems may then arise from the additive loss, the additive crosstalk, the polarization-dispersion loss, the polarization-mode dispersion, and from the decrease of channel spectral width in each unit.

The ability of different OADMs with four channels at 400-GHz spacing to be cascaded was investigated in a recirculating loop with optical amplifiers within the European research program ACTS [208, 209]. Devices based on multilayer and FP filters, bulk grating, fiber grating, acousto-optic filters, SiO2-Si MZI filters with grating, InP devices, either built in the ACTS (METON) project or commercial were tested. In most cases, more than 10 OADMs could be cascaded. Among the commercial components, the bulk grating OADM gave the best crosstalk performances: < −60 dB inter-channel and < −55 dB intrachannel crosstalks with total in/out loss of 7 to 9 dB. The fiber grating OADM gave the lowest loss: total in/out loss of 4 to 5 dB with < −50dB interchannel and < −25dB intrachannel crosstalks.

Very promising results have been obtained with silica on silicon waveguide grating filters. In [210], OADM with flattened passband MZIs incorporating FBGs, and OADM with AWGs were compared in a 4-wavelength channel transmission at 10 Gbps over 400 km through four OAMD nodes. In this configuration the losses are compensated using optical add/drop switches that have an amplifying function: Two cascaded erbium-doped fiber amplifiers are used in each OAMD. The comparison of the author is reported in Table 6.6.

Table 6.6
Comparison of Two Types of OAMD

	Mach-Zehnder/Bragg	AWG
Channel spacing	100 GHz	100 GHz
Maximum insertion loss	5.5 dB	7.0 dB
3-dB bandwidth	0.49 nm	0.33 nm
Minimum channel isolation	30 dB	22 dB
Typical thermal drift	0.001 nm/°C	0.01 nm

From: [210].

With the components available in this experiment, it is evident that the OADM using MZIs incorporating FBG are much better for cascading.

They have a larger FWHM, a better isolation; moreover, the thermal drift is smaller. After four nodes the power penalty with these devices was less than half that with the AWG devices. However, other devices with much better crosstalk and drift figures exist, especially AWG and free-space grating devices (down −55 dB/−60 dB crosstalk and 0.0005 nm/°C passive drift for the latter). Consequently, the technological choice is rather complex. Many aspects need to be evaluated in the design of a specific network. Of course, there is not a unique answer for all networks.

6.9.8 Optical Time Division Multiplexing Add/Drop

In this chapter we have presented up to this point only wavelength add/dropping. Fast interferometric add/drop switching for optical time division multiplexing (OTDM) is also feasible but will not be discussed. Semiconductor amplifiers have been widely investigated for this application. An example and more references are given in [211].

References

[1] Rose, D. L., "From SONET to Optical Networks," *NFOEC'99, Tech. Proc.*, Vol. 2, Chicago, Sept. 26–27, 1999, pp. 314–322.

[2] Chlamtac, I., A. Ganz, and G. Karmi, "Purely Optical Networks for Terabit Communication," *IEEE INFOCOM'89 Proc.*, 1989.

[3] Brackett, C. A., et al., "A Scalable Multiwavelength Multihop Optical Network: A Proposal for Research on All-Optical Networks," *IEEE J. Lightwave Technol.*, Vol. 11, May/June 1993.

[4] Hill, G. R., et al., "Special Issue on Broadband Networks," *IEEE J. Lightwave Technol.*, Vol. 11, May/June 1993.

[5] Alexander, S. B., et al., "Precompetitive Consortium on Wide-Band All Optical Networks," *IEEE J. Lightwave Technol.*, Vol. 11, May/June 1993.

[6] Houghton, A., "Research and Development of Optical Networks in the ACTS Program," *OPNET' 98 Proc.*, Paris, 1998.

[7] Jourdan, A., "L'Optique et les Télécommunications," *Actes Deuxième Entretien de la Physique*, Paris, Oct. 9–10, 1997, pp. 69–80.

[8] Srinivasan, N. V., "Multiwavelength Cross-Connect Technology," *LEOS'97 Proc.*, Vol. 2, San Francisco, CA, Nov. 10–13, 1997, abstract pp. 541.

[9] Hofmeister, R. T., et al, "Project LEARN: Light Exchangeable Add/Drop Ring Network," *OFC'97, Supplement to Technical Digest,* Post-deadline papers PD25, Dallas, TX, Feb. 16–21, 1997, pp. 1–4.

[10] Bentivoglio, F., and E. Iannone, "The Opaque Optical Network," *Optical Networks Magazine SPIE,* Vol. 1, No. 4, Oct. 2000, pp. 24–31.

[11] Bannister, J., M. Gerla, and M. Kovacevic, "An All-Optical Multi-Fiber Tree Network," *IEEE J. Lightwave Technol.*, Vol. 11, May/June 1993.

[12] Vodhanel, R. S., et al., "National-Scale WDM Networking Demonstration by the MONET Consortium," *OFC'97, Supplement to Technical Digest,* Post-deadline papers PD27, Dallas, TX, Feb. 16–21, 1997, pp. 1–4.

[13] Simmons, J., E. Goldstein, and A. Saleh, "On the Value of Wavelength-Add/Drop in WDM Rings with Uniform Traffics," *OFC'98 Proc.*, San Jose, CA, February 1998.

[14] Chiu, A. and E. Modiano, "Traffic Grooming Algorithms for Reducing Electronic Multiplexing Costs in WDM Ring Networks," *IEEE J. Lightwave Technol.*, January 2000.

[15] Chiu, A. L., and E. H. Modiano, "Reducing Electronic Multiplexing Costs in Unidirectional SONET/WDM Ring Networks Via Efficient Traffic Grooming," *Globecom'98 Proc.*, Sydney, Australia, Nov. 1998.

[16] Gerstel, O., P. Lin, and G. Sasaki, "Wavelength Assignment in a WDM Ring to Minimize Cost of Embedded SONET Rings," *Infocom '98 Proc.*, San Francisco, CA, March 1998.

[17] Zhang, X., and C. Qiao, "An Effective and Comprehensive Solution to Traffic Grooming and Wavelength Assignment in SONET/WDM Rings," *Conf. on All-Optical Networking, SPIE Proc.* Vol. 3531, Boston, Sept. 1998.

[18] Modiano, E., and A. Narula-Tam, "Mechanisms for Providing Optical Bypass in WDM-Based Networks," *Optical Networks*, Vol.1, No. 1, Jan. 2000, pp. 11–18.

[19] Russ, W., "Optical Switching Reduces Cost of Internet Transport," *FiberOptic Product News*, Vol. 4, April 2000, pp. 31–34.

[20] Grinsec, M., *La Commutation Électronique*, CNET-ENST, Paris: Eyrolles Ed., 1980.

[21] Brackett, C., "Dense Wavelength Division Multiplexing Networks: Principles and Applications," *IEEE J. of Selected Areas in Comm.*, Vol. 8, August 1990, pp. 948–964.

[22] Kobayashi, H., "Duality Relationships Among 'Space,' 'Time,' and 'Wavelength' in All-Optical Networks," *IEEE J. Lightwave Technol.*, Vol. 14, No. 3, 1996, pp. 344–351.

[23] Beauquier, B., "All-to-All Communication for Some Wavelength-Routed All-Optical Networks," *Networks (NY)*, Vol. 33, No. 3, 1999, pp. 179–187.

[24] Mestdagh, D. J., *Fundamental of Multi-Access Optical Fiber Networks*, Norwood, MA: Artech House, 1995.

[25] Lagasse, P., et al., "Photonics Technologies in Europe," *ACTS HORIZON AC 058*, European Commission Research Programme, http://www.infowin.org and http://intec.rug.as.be.horizon, 2000.

[26] Ramamurthy, B., and B. Mukherjee, "Wavelength Conversion in WDM Networking," *IEEE J. of Selected Areas in Comm.*, Vol. 16, No. 7, Sept. 1998, pp. 1061–1073.

[27] Späth, J., and S. Bodamer, "Performance Evaluation of Photonic Networks Under Dynamic Traffic Conditions," *Proc. 2nd IFIC TC6 Working Conf. On Optical Network Design and Modeling*, Rome, Feb. 1998, pp. 15–19.

[28] Späth J., and S. Bodamer, "Routing Strategies for Photonic Networks Under Dynamic Traffic Conditions," *NOC'98 Proc.*, 1998.

[29] Späth, J., and S. Bodamer, "Routing of Dynamic Poisson and Non-Poisson Traffic in WDM Networks with Limited Wavelength Conversion," *ECOC'98 Proc.*, Vol. 1, Madrid, Sept. 20–24, 1998, pp. 359–360.

[30] Baroni, S., and P. Bayvel, "Wavelength Requirements in Arbitrarily Connected Wavelength-Routed Optical Networks," *IEEE/OSA, IEEE J. Lightwave Technol.*, Vol. 15, Feb. 1997, pp. 242–251.

[31] Barry, R. A. and D. Marquis, "An Improved Model of Blocking Probability in All-Optical Networks," *Proc. IEEE LEOS'95 Summer Topical Meeting*, 1995, pp. 43–44.

[32] Subramamian, S., M. Azizoglu, and A. K. Somani, "Connectivity and Space Wavelength Conversion in Wavelength-Routed Networks," *Proc. IEEE INFOCOM'96*, 1996, pp. 148–155.

[33] Jeong, G., and E. Ayanoglu, "Comparison of Wavelength Interchanging and Wavelength-Selective Cross-Connects in Multiwavelength All Optical Networks," *IEEE INFOCOM'96 Proc.*, 1996, pp. 156–163.

[34] Garnot, M., et al., "Wavelength Conversion in Future WDM Optical Transport Networks," *Proc. IEEE LEOS'96 Summer Topical Meeting*, 1996, pp. 47–48.

[35] Flammini, M., and C. Scheideler, "Simple, Efficient Routing Schemes for All-Optical Networks," *Theory Compt. Systems*, Vol. 32, No. 3, 1999, pp. 387–420.

[36] Wagner, R. E., et al., "Realizing the Vision of Multiwavelength Networking," *IOOC-ECOC'97 Proc.*, Vol. 3, No. 448, IEE Ed., Edinburgh, Sept. 22–25, 1997, pp. 143–147.

[37] Karasan, E., and E. Ayanoglu, "Performance Comparison of Reconfigurable Wavelength-Selective and Wavelength-Interchanging Cross-Connects in WDM Transport Networks," *NFOEC'97 Proc.*, Paper 21.4, San Diego, CA, Sept. 21–27, 1997.

[38] Baroni S., S. K. Korotky, and P. Bayvel, "Wavelength Interchange in Multi-Wavelength Optical Transport Networks," *Proc. IOOC-ECOC'97*, Vol. 3, No. 448, IEE Ed., Edinburgh, Sept. 22–25, 1997, pp. 164–167.

[39] Clos, C., "A Study of Nonblocking Switching Networks," *Bell Syst. Tech. J.*, Vol. 32, March 1953, pp. 406–424.

[40] Hui, J. Y., *Switching and Traffic Theory for Integrated Broadband Networks*, Boston: Kluwer, 1990.

[41] Kobayashi, H., B. L. Mark, and Y. Osaki, "Call Blocking Probability of All-Optical Networks," *Proc. 1st IEEE Int. Workshop Broadband Switching Syst. Poznan*, Poland, April 1995, pp. 186–200.

[42] Brackett, C. A., "Foreword: Is There an Emerging Consensus on WDM Networking?," *IEEE J. Lightwave Technol.*, Vol. 14, No. 6, 1996, pp. 936–941.

[43] Bala, K., F. Chung, and C. Brackett, "Optical Wavelength Routing, Translation, and Packet/Cell Switched Networks," *IEEE J. Lightwave Technol.*, Vol. 14, No. 3, 1996, pp. 336–343.

[44] Chang, C. E., et al, "40 Gbps WDM Cross-Connect with an Electronic Switching Core: Preliminary Results from the WEST Consortium," *LEOS'97 Proc.*, Vol. 2, San Francisco, Nov. 10–13, 1997, pp. 336–337.

[45] Kobrinski, H., "Crossconnection of Wavelength Division Multiplexed High Speed Channels," *Electron. Lett.*, Vol. 23, 27 August 1987, pp. 975–977.

[46] Wagner, S. S., and H. Kobrinski, "WDM Applications in Broadband Telecommunication Networks," *IEEE Communications Magazine*, March 1989, pp. 22–30.

[47] Karasan, E., and E. Ayanoglu, "Performance of WDM Transport Networks," *IEEE J. of Selected Areas in Comm.*, Vol. 16, No. 7, Sept. 1998, pp. 1081–1096.

[48] Neefs, H., "Optoelectronic Interconnects for Integrated Circuits: Achievements 1996–2000," *Advanced Research Initiative in MicroElectronics MEL-ARI OPTO*, Office for Official Publications of the European Communities, Luxemburg, June 2000, p. 13.

[49] Palissery, R., "True Interoperability in Optical Networks," *NFOEC'99, Tech. Proc.*, Vol. 1, Chicago, Sept. 26–27, 1999, pp. 516–525.

[50] Popelek, J., and Y Li, "Free-Space-Fiber Hybrid Distributed Optical Cross-Connect InterConnect Module," *Optics Lett.*, Vol. 24, No. 3, Feb. 1999, pp. 142–144.

[51] Benabes, P. A., et al., "A Terabit/s 0.6 μm CMOS InterConnect Cross-Bar for Optoelectronic Communication System," *Proc. International Conf. on Electronics Circuits and Systems, ICECS'99*, Paphos, Cyprus, Sept. 5–8, 1999, pp. 257–260.

[52] Fancey, S. J., et al., "A Free-Space Optoelectronic Cross-Bar InterConnect with Terabit/s Communication to Silicon Electronics," *Proc. CLEO Europe'98*, Glasgow, Scotland, Sept. 1998.

[53] Walker, A. C., et al., "Design and Construction of an Optoelectronic Cross-Bar Switch Containing a Terabit/s Tree-Space Optical InterConnect," *IEEE J. Selected Topics in Quant. Electron.*, Vol. 5, No. 2, 1999, pp. 236–249.

[54] Neefs, H., "Optoelectronic InterConnects for Integrated Circuits: Achievements 1996–2000," *Advanced Research Initiative in MicroElectronics MEL-ARI OPTO*, Office for Official Publications of the European Communities, Luxemburg, June 2000, pp. 20–23.

[55] Gillner, L., and M. Gustavsson, "Expansion of Amplifier Gate Switch Array Size in Optical Multiwavelength Switching Networks," *ECOC'94 Proc.*, Vol. 2, Florence, Italy, Sept. 1994, pp. 537–543.

[56] Hadjifotiou, A., "Network Applications for Optical Cross-Connect," *WDM Conference Reprints*, London: Vision in Business Ed., Nov. 1997.

[57] Fainman, Y., "Optical Interconnect Systems for Communication and Computing" *LEOS'97 Proc.*, Vol. 2, San Francisco, CA, Nov. 10–13, 1997, pp. 343–344.

[58] Perrier, P. A., "Position, Functions, Features, and Enabling Technologies of Optical Cross-Connects in the Photonic Layer," *NFOEC'99, Tech. Proc.*, Vol. 1, Chicago, Sept. 26–27, 1999, pp. 260–269.

[59] Lee, B. H., and R. J. Capik, "Demonstration of a Very Low-Loss, 576 × 576 Servo-Controlled, Beam-Steering Optical Switch Fabric," *ECOC'2000 Proc.*, Vol. 4, Munich, Sept. 3–5, 2000, pp. 95–96.

[60] Laor, H., et al., "Performance of a 576 × 576 Optical Cross Connect," *NFOEC'99, Tech. Proc.*, Vol. 1, Chicago, Sept. 26–27, 1999, pp. 276–281.

[61] Wu, M., "Micromachining for Optical and Optoelectronic Systems," *Proc. IEEE*, Vol. 85, No. 11, Nov. 1997, pp. 1833–1856.

[62] Hornbeck, L. J., "Current Status of the Digital Micromirror Device (DMD) for Projection Television Applications," *International Electron Devices Technical Digest*, 1993, pp. 381–384.

[63] Asksyuk, V., et al., "Low Insertion Loss Packaged and Fiber Connectorized Si Surface-Micromachined Reflective Optical Switch," *Solid-State Sensor and Actuator Workshop Proc*, Hilton Head Island, SC, June 8–11, 1998., pp. 79–82.

[64] Giles, R., et al., "Highly Efficient Light-Actuated Micromechanical Photonic Switch for Enhanced Functionality at Remote Nodes," *OFC'98 Proc.*, Post-deadline paper PD 2, San Jose, CA, 1998.

[65] Ford, J. E., et al., "Wavelength-Selectable Add/Drop with Tilting Micromirrors," *LEOS Annual Meeting Proc.*, Post-deadline paper PD 2.3, Nov. 1997.

[66] Randy, G. C., "LightWave Micromachines," *ECOC'98 Proc.*, Vol. 1, Madrid, Sept. 20–24, 1998, pp. 249–250.

[67] Lin, L. Y., E. L. Goldstein, and R. W. Tkach, "Free-Space Micromechanical Optical CrossConnect with Bridging Functionality for Optical-Layer Restoration," *ECOC'98 Proc.*, Vol. 1, Madrid, Sept. 20–24, 1998, pp. 459–460.

[68] Lin, L. Y., and E. L. Goldstein, "MEMS for Optical Networking: Current Status and Future Prospects," *ECOC'2000 Proc.*, Vol. 4, Munich, Sept. 3–5, 2000, pp. 91–92.

[69] Lin, L. Y., E. L. Goldstein, and R. W. Tkach, "Free-Space Micromachined Optical Switches with Sub-Millisecond Switching Time for Large-Scale Optical Cross-Connects," *IEEE Photon. Technol. Lett.*, Vol. 10, No. 4, 1998, pp. 525–527.

[70] Reed, J., "An Update on the Latest Technological Developments Supporting an Integrated Optically Enhanced Network," *WDM Conference Reprints*, London: Vision in Business Ed., Nov. 1997.

[71] Keil, N., et al., "Thermo-Optic Switches Using Vertically Coupled Polymer/Silica Waveguides," *ECOC'2000 Proc.*, Vol. 4, Munich, Sept. 3–5, 2000, pp. 101–102.

[72] Keil, N., et al., "Polymer Waveguide Optical Switch with –40 dB Polarisation Independant Crosstalk," *Electron. Lett.*, Vol. 32, No. 7, 1996, pp. 655–657.

[73] Hauffe, R., U. Siebel, and K. Petermann, "Crosstalk Minimization in Integrated Switching Matrixes," *ECOC'2000 Proc.*, Vol. 4, Munich, Sept. 3–5, 2000, pp. 97–98.

[74] Michiyuki, A., et al., "Digital Thermo-Optic Switch," *NTT Opto-Electronics Laboratories Annual Report 97*, 1997, p. 44.

[75] Michiyuki, A., et al., "Low Loss and High-Extinction Ratio Thermo-Optic 8 × 8 Matrix Switch Using Silica-Based PLCs," *NTT Opto-Electronics Laboratories Annual Report 97*, 1997, p. 35.

[76] Michiyuki, A., et al., "Self-Latching Optical Waveguide Switch Based on Thermo-Capillarity," *NTT Opto-Electronics Laboratories Annual Report 97*, 1997, p. 4.

[77] Nakajima, I., et al., "Prototype WP-Based Optical Path Cross-Connect Node Using PI-Loss Optical Switches," *ECOC'98 Proc.*, Vol. 1, Madrid, Sept. 20–24, 1998, pp. 251–252.

[78] Liu, J. Y., and S. Grout, "New Technologies for Active Optical Networking," *NFOEC'99, Tech. Proc.*, Vol. 1, Chicago, Sept. 26–27, 1999, pp. 270–275.

[79] Yamazaki, H., et al., "Large-Scale Holographic Switch with a Ferroelectric Liquid-Crystal Spatial Modulator," *LEOS'97 Proc.*, Vol. 1, San Francisco, CA, Nov. 10–13, 1997, pp. 128–129.

[80] Wolffer, N., et al., "8 × 8 Holographic Liquid Crystal Switch," *ECOC'2000 Proc.*, Vol. 3, Munich, Sept. 3–5, 2000, pp. 275–276.

[81] Hironishi, K., et al., "4 × 4 PI-LOSS-Topology Based Free-Space Switch for Optical Cross-Connect Systems," *ECOC'95 Proc.*, Vol. 1, Brussels, Sept. 17, 1995, pp. 123–126.

[82] Noguchi, K., and W. Kawakami, "Optical Digital Cross-Connect System for Broadband WDM Network Reconfiguration Using Liquid-Crystal Optical Multichannel Switches," *ECOC'95 Proc.*, Vol. 1, Brussels, Sept. 17, 1995, pp. 127–130.

[83] Wu, K. -Y., and J. -Y Liu, "Liquid-Crystal Space and Wavelength Routing Switches," *LEOS'97 Proc.*, Vol. 1, San Francisco, CA, Nov. 10–13, 1997, pp. 28–29.

[84] Khan, M. N., et al., "Fabrication-Tolerant, Low-Loss, and High-Speed Digital Optical Switches in InGaAsP/InP Qantum Wells," *ECOC'95 Proc.*, Vol. 1, Brussels, Sept. 17, 1995, pp. 103–106.

[85] Renaud, M., et al., "Compact Digital Optical Switches for Low Insertion Loss Large Switch Arrays on InP," *ECOC'95 Proc.*, Vol. 1, Brussels, Sept. 17, 1995, pp. 99–102.

[86] Hoffmann, D., et al., "Novel Digital Optical Switch with Crosstalk Below –40 dB Based on Absorptive Switching," *ECOC'95 Proc.*, Vol. 1, Brussels, Sept. 17, 1995, pp. 107–110.

[87] Michiyuki, A., et al., "Electro-Optic Switch Constructed with a Poled Silica-Based Waveguide on a Si Substrate," *NTT Opto-Electronics Laboratories Annual Report 97*, 1997, p. 36.

[88] Murphy, T., et al., "A Strictly Non-Blocking 16 × 16 ElectroOptic Photonic Switch Module," *ECOC'2000 Proc.*, Vol. 4, Munich, Sept. 3–5, 2000, pp. 93–94.

[89] Gustavsson, M., et al., "Monolithically Integrated 4 × 4 InGaAsP/InP Laser Amplifier Gate Switch Arrays," *Electron. Lett.*, Vol. 28, No. 24, Nov. 19, 1992, pp. 2223–2225.

[90] Kalman, R. F., et al., "Space Division Switches Based on Semiconductor Optical Amplifiers," *IEEE Photon. Technol. Lett.*, Vol. 4, No. 9, Sept. 1992, pp. 1048–1051.

[91] Lach, E., et al., "InP-Based Space Switch Module for 1.55 μm with Wide Wavelength Span, Low Noise and 10 Gbit/s Transparency," *ECOC'95 Proc.*, Vol. 1, Brussels, Sept. 17, 1995, pp. 111–114.

[92] Gillner, L., "Comparison of Dynamic Effects in Two Types of Expanded Semiconductor Optical Amplifier Gate Switch Arrays," *ECOC'95 Proc.*, Vol. 1, Brussels, Sept. 17, 1995, pp. 119–122.

[93] Takeshita, H., et al., "A Demonstration of an Optical Cross-Connect System for a Self-Healing Optical Network," *IOOC-ECOC'97 Proc.*, Vol. 3, No. 448, IEE Ed., Edinburgh, Sept. 22–25, 1997, pp. 335–338.

[94] Varrazza, R., et al., "Demonstration of Simultaneous Packet Routing and Wavelength Conversion at 10Gb/s in a Highly Compact, Lossless Vertical Coupler Optical Switch," *ECOC'2000 Proc.*, Vol. 4, Munich, Sept. 3–5, 2000, pp. 67–68.

[95] Sartorius, B., "Photon Under Control: Optical Signal Processing for High-Speed Networks," *IST'2000 Proc., MicroElectronics, Optoelectronics and Networks Session Proc.*, Nice, France, Nov. 7, 2000, pp. 5–6.

[96] Dorgeuille, F., et al., "First Array of 8 CG-SOA Gates for Large-Scale WDM Space Switches," *ECOC'98 Proc.*, Vol. 1, Madrid, Sept. 20–24, 1998, pp. 255–256.

[97] Diez, S., et al., "Novel Gain-Transparent SOA-Switch for High Bitrate OTDM Add/Drop Multiplexing," *ECOC'98 Proc.*, Vol. 1, Madrid, Sept. 20–24, 1998, pp. 461–462.

[98] Harris, S. E., and R. W. Wallace, "AcoustoOptic Tunable Filter," *J. Opt. Soc. Am.*, Vol. 59, No. 6, 1969, pp. 744–747.

[99] Goto, N., and Y. Miyazaki, "Integrated Multi-Demultiplexer Using AcoustoOptic Effect for Multiwavelength Optical Communications," *IEEE J. on Selected Areas in Comm.*, Vol. 8, No. 6, August 1990, pp. 1160–1168.

[100] Cheung, K. W., et al. "Multiple Channel Operation of Integrated Acousto Optic Tunable Filter," *Electron. Lett.*, Vol. 25, No. 6, 16 March 1989, pp. 375–376.

[101] Cheung, K. W., et al., "1 Gbps System Performance of an Integrated, Polarization-Independant, Acoustically-Tunable Optical Filter," *IEEE Photon. Technol. Lett.*, Vol. 2, 1990, pp. 271–273.

[102] Cheung, K. W., et al., "Simultaneous Five Wavelength Filtering at 2.2 nm Wavelength Separation Using Integrated Acoustooptic Tunable Filter with Subcarrier Detection," *Electron. Lett.*, Vol. 25, No. 10, May 11, 1989, pp. 636–637.

[103] Kinoshita, T., and K. I. Sano, "Design and Performance of a Tunable Optical Demultiplexer Using an AcoustoOptic Light Deflector," *Elec. and Com. in Japan*, Vol. 72, Part 2, No. 6, 1989, pp. 14–22.

[104] Smith, D. A., et al., "Integrated Optic Acoustically Tunable Filters for WDM Networks," *IEEE J. on Selected Areas in Comm.*, Vol. 8, No. 6, August 1990.

[105] Heffner, B. L. , et al., "Integrated Optic Acoustically Tunable Infrared Optical Filter," *Electron. Lett.*, Vol. 24, No. 25, December 8, 1988, pp. 1562–1563.

[106] Shimazu, Y., and S. Kikuchi, "Time and Wavelength Division Optical Distribution System Using AcoustoOptic Tunable Filter," in "Fiber Networking and Telecommunications," Proc. *SPIE* 1179, 1989, pp. 34–42.

[107] Liew, S. C., "A Multiwavelength Optical Switch Based on the Acousto-Optic Tunable Filter," *SPIE*, Vol. 1363, 1990, pp. 5761.

[108] Riza, N. A., "Low Interchannel Crosstalk Wavelength Routing Switch Based on Bulk Acousto-Optic Tunable Filters," *LEOS'97 Proc.*, Vol. 2, San Francisco, CA, Nov. 10–13, 1997, pp. 341–342.

[109] Chang, I. C., "Collinear Beam Acousto-Optic Tunable Filters," *Electron. Lett. (U.K.)*, Vol. 28, No. 13, June 1992, pp. 1255–1256.

[110] Cheung, K. W., "AcoustoOptic Tunable Filters in Narrowband WDM Network Applications," *IEEE J. Select. Areas Comm.*, Vol. 8, 1990, pp. 1015–1025.

[111] Wehrmann, F., et al., "Integrated Optical Wavelength Selective Acoustically Tunable 2x2 Switches (Add/Drop Multiplexers) in $LiNbO_3$," *IEEE J. Select. Topics Quantum Electron.*, Vol. 2, 1996, pp. 263–269.

[112] Smith, D., et al., "Reduction of Crosstalk in an Acousto-Optic Switch by Means of Dilatation," *Opt. Lett.*, Vol. 19, 1994, pp. 99–101.

[113] Hermann, H., et al., "Double Stage, Integrated, Acousto-Optical Add/Drop Multiplexers with Improved Crosstalk Performance," *IOOC-ECOC'97 Proc.*, Vol. 3, No. 448, IEE Ed., Edinburgh, Sept. 22–25, 1997, pp. 10–13.

[114] Kobrinski, H., "Cross-Connection of Wavelength-Division-Multiplexed High-Speed Channels," *Electron. Lett.*, Vol. 23, Aug. 27, 1987, pp. 975–977.

[115] Hill, G. R., "Wavelength Routing Approach to Optical Communication Networks," *Infocom'88 Proc.*, New Orleans, LA, March 1988.

[116] Wagner, S. S., and H. Kobrinski, "WDM Applications in Broadband Telecommunication Networks," *IEEE Communications Magazine*, March 1989, pp. 22–30.

[117] Sabella, R., "Tutorial: Key Elements for WDM Transport Networks," *Photonic Network Com.*, Vol. 2, No. 1, Jan./Mar. 2000, pp. 7–13.

[118] Wagner, R. E., "Multiwavelength Optical Networks," *ECOC'95 Proc.*, SC2 Tutorial, Brussels, Sept. 17, 1995.

[119] Laude, J. P., *Wavelength Division Multiplexing*, London: Prentice Hall, 1993, pp. 105–110.

[120] Banerjee, D., J. Frank, and B. Mukherjee, "Passive Optical Network Architecture Based on Waveguide Grating Routers," *IEEE J. of Selected Areas in Comm.*, Vol. 16, No. 7, Sept. 1998, pp. 1040–1050.

[121] Frigo, N. J., et al., "A Wavelength Division Multiplexed Passive Optical Network with Cost-Shared Components," *IEEE Photon. Technol. Lett.*, Vol. 6, 1994, pp. 1265–1367.

[122] Iannone, P. P., N. J. Frigo, and T. E. Darcie, "A WDM PON Architecture with Bidirectional Optical Spectral Splicing," *OFC'95 Technical Digest*, Vol. 8, San Diego, CA, Feb. 1995, pp. 51–53.

[123] Zirngibl, M., et al., "LAR-Net, a Local Access Router Network, " *IEEE Photon. Technol. Lett.*, Vol. 6, 1995, pp. 215–217.

[124] Hill, A. et al., "Multiple-Star Wavelength-Router Network and its Protection Strategy," *IEEE J. of Selected Areas in Comm.*, Vol. 16, No. 7, Sept. 1998, pp. 1134–1145.

[125] Kato, K., et al., "10-Tbps Full-Mesh WDM Network Based on Cyclic-Frequency Arrayed-Waveguide Grating Router," *ECOC'2000 Proc.*, Vol. 1, Munich, Sept. 3–5, 2000, pp. 105–106.

[126] Sato, T., and T. Horiguchi, "A Novel Bi-Directional Optical Amplifier Using AWGs for Dense WDM Systems," *ECOC'98 Proc.*, Madrid, Sept. 20–24, 1998, pp. 635–636.

[127] Laude, J. P., I. Long, and D. Fessard, "Very Dense NxN Wavelength Routers Based on a New Diffraction Grating Configuration," *IOOC-ECOC'97 Proc.*, Vol. 3, No. 448, IEE Ed., Edinburgh, Sept. 22–25, 1997, pp. 87–90.

[128] Laude, J. P., I. Long, and D. Fessard, "Low Loss, Low Crosstalk MxN Passive Wavelength Routers Based on Diffraction Gratings," *LEOS'97 Proc.*, O*A*Vol. 2, San Francisco, CA, Nov. 10–13, 1997, pp. 506–507.

[129] Laude, J. P., and K. Lange, "Dense Wavelength Division Multiplexers and Routers Using Diffraction Gratings," *NFOEC'99, Tech. Proc.* Vol. 1, Chicago, Sept. 26–27, 1999, pp. 83–88.

[130] Churin, E. G., and P. Bayvel, "Free-Space, Dense WDM Router Based on a New Concave Grating Configuration," *ECOC'98 Proc.*, Vol. 1, Madrid, Sept. 20–24, 1998, pp. 239–240.

[131] Stavdas, A., et al., "Design and Performance of Concave Holographic Gratings for Applications as Multiplexers/Demultiplexers for Wavelength Routed Optical Networks," *Opt. Eng.*, Vol. 35, 1996, p. 2816.

[132] Smit, M. K., "New Focusing and Dispersive Planar Component Based on Optical Phase Array," *Electron. Lett.*, Vol. 24, 1988, pp. 385–386.

[133] Takahashi, H., et al., "Arrayed-Waveguide Grating for Wavelength Division Multi/Demultiplexer with Nanometer Resolution," *Electron. Lett.*, Vol. 26, 1990, pp. 87–88.

[134] Dragone, C., "An NxN Optical Multiplexer Using a Planar Arrangement of Two Star Couplers," *IEEE Photon. Technol. Lett.*, Vol. 3, 1991, pp. 812–815.

[135] Okamoto, K., K. Moriwaki, and S. Suzuki, "Fabrication of 64 × 64 Arrayed-Waveguide Grating on Silicon," *Electron. Lett.*, Vol. 31, 1995, pp. 184–186.

[136] Okamoto, K. et al., "32 × 32 Arrayed-Waveguide Grating Multiplexer with Uniform Loss and Cyclic Frequency Characteristics," *Electron. Lett.*, Vol. 33, 1997, pp. 1865–1866.

[137] Maru, K., et al., "16 × 32 AWG with Cyclic-Frequency Response," *OECC'98 Proc.*, Chiba, Japan, July 12–16, 1998, pp. 54–55.

[138] Harada, K., et al., "Hierarchical Optical Path Cross-Connect Systems for Large Scale WDM Networks," *OFC'99 Proc.*, poster WM55, 1999.

[139] Blaizot, C., et al., "Multi-Granularity Optical Networks," *ONDM'2000 Proc.*, Athens, Jan. 2000, p.1.

[140] Noirie, L., C. Blaizot, and E. Dotaro, "Multi-Granularity Optical Cross-Connect," *ECOC'2000 Proc.*, Vol. 3, Munich, Sept. 3–5, 2000, pp. 269–270.

[141] Ciaramella, E., "Bundling Wavelengths in Optical Transport Networks," *ECOC'2000 Proc.*, Vol. 3, Munich, Sept. 3–5, 2000, pp. 235–236.

[142] Bernasconi, P., et al., "Large N × N Waveguide Grating Routers," *IEEE J. of Lightwave Technol.*, Vol. 18, No. 7, July 2000, pp. 985–991.

[143] Gasman, L. D., et al., "*Wave Division Multiplexing, Photonic Switching and the Coming of All Optical Networks*," Vol. 1, 1999–2000, Com. Industry Researchers, Inc., Charlottesville, NC, 2000.

[144] Okamoto, K., et al., "16-Channel Optical Add-Drop Multiplexer Consisting of Arrayed Waveguide Gratings and Double-Gate Switches," *Electron. Lett.*, Vol. 32, No. 16, 1996, p. 1471.

[145] Hemenway, B. R., et al., "Demonstration of a Re-Configurable Wavelength-Routed Network at 1.14 Terabits-Per-Second," *OFC'97, Supplement to Technical Digest*, Post-deadline papers PD26, Dallas, TX, Feb. 16–21, 1997, pp. 1–4.

[146] Maeno, Y., et al., "10 Gbit/port and 256 Ports Optical CrossBar Switch with Reduced Hardware and Enhanced Modularity," *IOOC-ECOC'97 Proc.*, Vol. 3, No 448, IEE Ed., Edinburgh, Sept. 22–25, 1997, pp. 67–70.

[147] Takahashi, H., K. Okamoto, and Y. Inoue, "Arrayed-Waveguide Grating Wavelength Multiplexers for WDM Systems," *NTT Review*, Vol. 10, No. 1, Jan. 1998, pp. 37–44.

[148] Tachikawa, Y., et al., "Optical Add-Drop Multiplexers and Multi-Wavelength Optical Sources for WDM Transmission Systems," *NTT Review*, Vol. 10, No. 1, Jan. 1998, pp. 52–59.

[149] Herben, C. G. P., et al., "Compact Integrated Polarization-Independent Optical Cross-Connect," *ECOC'98 Proc.*, Vol. 1, Madrid, Sept. 20–24, 1998, pp. 257–258.

[150] Herben, C. G. P., et al., "A Compact Integrated InP-Based Single-PHASAR Optical Cross-Connect," *IEEE Photon. Technol. Lett.*, Vol. 10, No. 5, May 1998.

[151] Farjady, F., M. C. Parker, and S. D. Walker, "High Usage Optical Access Architecture Featuring Coarse Space-Wavelength Routing," *ECOC'98 Proc.*, Vol. 1, Madrid, Sept. 20–24, 1998, pp. 581–582.

[152] Smit, M., "Advanced Devices for WDM Applications," *ECOC'2000, Tutorial 4*, Munich, Sept. 3–5, 2000.

[153] Liaw, S. -K., S. Chi, and K. -P Ho, "Experimental Investigation of the Critical Issues on Optical Cross-Connect Devices Using Fiber Bragg Gratings," *J. of Opt. Com.*, Vol. 21, August 2000, pp. 131–133.

[154] Koonen, T., et al., "Wavelength Routing in Fiber-Wireless Networks with Spectral Selection and Remote Modulation," *ECOC'2000 Proc.*, Vol. 1, Munich, Sept. 3–5, 2000, pp. 33–34.

[155] Sharony, J., K. W. Cheug, and T. E. Stern, "The Wavelength Dilatation Concept-Implementation and System Considerations," *ICC'92 Proc.*, Paper 330.3, Chicago, June 1992.

[156] Sharony, J., K. W. Cheung, and T. E. Stern, "Wavelength Dilated Switches (WDS), A New Class of High Density, Suppressed Crosstalk, Dynamic Wavelength-Routing CrossConnects," *IEEE Photon. Technol. Lett.*, Vol. 4, No. 8, August 1992, pp. 933–935.

[157] Shah, V., et al., "10 Gb/s Transmission Across 4 Optical Cross-Connect Nodes and 200 km Fiber," *ECOC'98 Proc.*, Vol. 1, Madrid, Sept. 20–24, 1998, pp. 601–602.

[158] Holmstrom, R. P., and L. Wosinska, "Improving the Reliability of Optical Cross-Connects by Using Shared Protection," *NFOEC'99, Tech. Proc.*, Vol. 2, Chicago, Sept. 26–27, 1999, pp. 592–596.

[159] Kong, E., et al., "A Novel Optical Cross-Connect with Built-In Optical-Path Supervisory Scheme for All-Optical Networks," *ECOC'98 Proc.*, Vol. 1, Madrid, Sept. 20–24, 1998, pp. 585–586.

[160] Louis, S., et al., "Supervisory Channels in an Optical Demultiplexer," *WFOPC'98, LEOS-IEEE Proc.*, Pavia, Italy, 18–19 Sept. 1998, pp. 15–18.

[161] Bersiner, L., et al., "Experimental Optical CrossConnect System with Automatic Protection Switching for Wavelength Division Multiplex Networks," *ECOC'98 Proc.*, Vol. 1, Madrid, Sept. 20–24, 1998, pp. 615–616.

[162] Hudendick, S., and J. Söderqvist, "Management of Connection Services in Dynamic Optical Networks," *ECOC'2000 Proc.*, Vol. 3, Munich, Sept. 3–5, 2000, pp. 237–238.

[163] Harai, H., M. Murata, and H. Miyahara, "Performance Analysis of Wavelength Assignment Policies in All-Optical Networks with Limited-Range of Wavelength Conversion," *IEEE J. of Selected Areas in Comm.*, Vol. 16, No. 7, Sept. 1998, pp. 1051–1060.

[164] Bunse, S., et al., "Concepts and Components for Optical Routing Nodes in Transparent and Packet Based High Speed Networks," *ECOC'95 Proc.*, Vol. 1, Brussels, Sept. 17, 1995, pp. 429– 432.

[165] Schilling, M., et al., "Monolithic Mach-Zehnder Interferometer Based Optical Wavelength Converter Operated at 2.5 Gbit/s with Extinction Ratio Improvement and Low Penalty," *ECOC'94 Proc.*, 1994, pp. 647–650.

[166] StaRing, A., et al., "Space-Switching 2.5 Gbit/s Signals Using Wavelength Conversion and Phased Array Routing," *Electron. Lett.*, Vol. 32, No. 4, Feb. 15, 1996, pp. 377–379.

[167] Zouganeli, E., et al., "Wavelength Routed Network Using Widely Tunable Laser Transmitters," *ECOC'2000 Proc.*, Vol. 4, Munich, Sept. 3–5, 2000, pp. 51–52.

[168] Broberg, B., et al., "Widely Tunable Semiconductor Lasers," *OFC/IOOC'99 Proc.*, Invited paper, San Diego, CA, 1999.

[169] Robbins, D. J., "Design and Optimization of Sampled Grating DBR Laser for Dense WDM Networks," *ECOC'98 Proc.*, Vol. 1, 1998, pp. 221–222.

[170] Ishii, H., et al., "Quasicontinuous Wavelength Tuning in Super-Structure (SSG) DBR Lasers," *IEEE J. Quantum Electronics*, Vol. 32, 1996, pp. 433–440.

[171] Kuroyanagi S., and T. Maeda, "An Optical Cross-Connect Architecture Incorporating Failure Recovery Using Reserved Wavelengths," *Proc. Photonics in Switching Conf.*, Vol.12, Salt Lake City, UT, 1995.

[172] Sabella, R., "Tutorial: Key Elements for WDM Transport Networks," *Photonic Network Com.*, Vol. 2, No 1, Jan./Mar. 2000, pp. 7–13.

[173] Tomkos, I., D. Syvridis, and N. Antoniades, "Modular and Re-Configurable Parametric Wavelength Interchanging Cross-Connect Architectures," *Photonic Network Com.*, Vol. 2, No 1, Jan./Mar. 2000, pp. 53–59.

[174] Karasan, E., and E. Ayanoglu, "Performance of WDM Transport Networks," *IEEE J. of Selected Areas in Comm.*, Vol. 16, No. 7, Sept. 1998, pp. 1081–1096.

[175] Teshima, M., et al., "Demonstration of Virtual Wavelength Path Cross-Connect," *IOOC-ECOC'97 Proc.*, Vol. 3, No. 448, IEE Ed., Edinburgh, Sept. 22–25, 1997, pp. 59–62.

[176] Ishida, O., N. Takachio, and K. -I. Sato, "Modular Cross-Connect System for WDM Optical-Path Networks," *IOOC-ECOC'97 Proc.*, Vol. 3, No. 448, IEE Ed., Edinburgh, Sept. 22–25, 1997, pp. 63–66.

[177] Iannone, E., and F. Bentiviglio, "Innovative Multi-Modular Architecture Design for an Optical Cross-Connect: A Present Perspective," *NFOEC'99 Tech. Proc.*, Vol. 2, Chicago, Sept. 26–27, 1999, pp. 697–707.

[178] Jourdan A., "L'Optique et les Télécommunications," *Les 2e Entretiens de la Physique, Ed. SFP- CNISF*, Paris, Oct. 9–10, 1997, pp. 69–80.

[179] Jourdan, A., et al., "Fully Reconfigurable WDM Optical CrossConnect: Feasibility Validation and Preparation of Prototype CrossConnect for ACTS 'OPEN' Field Trials," *IOOC-ECOC'97 Proc.*, Vol. 3, No. 448, IEE Ed., Edinburgh, Sept. 22–25, 1997, pp. 55–58.

[180] Lin, L. Y., E. Karasan, and R. Tkach, "Layered Switch Architecture for High-Capacity Optical Transport Networks," *IEEE J. of Selected Areas in Comm.*, Vol. 16, No. 7, Sept. 1998, pp. 1074–1080.

[181] Goldstein E. L., L. Eskilden, and A. F. Elrefaie, "Performance Implications of Components Crosstalk in Transparent Lightwave Networks," *IEEE Photon. Technol. Lett.*, Vol. 6, May 1994, pp. 657–660.

[182] Li, C. S., and F. Tong, "Crosstalk and Interference Penalty in All-Optical Networks Using Static Wavelength Routers," *IEEE J. Lightwave Technol.*, Vol. 14, June 1996, pp. 1120–1126.

[183] Takahashi, H., K. Oda, and H. Toba, "Impact of Crosstalk in an Arrayed Waveguide Multiplexer on NxN Optical InterConnections," *IEEE J. Lightwave Technol.*, Vol. 14, June 1996, pp. 1097–1105.

[184] Antoniade, N., et al., "Crosstalk Performance of a Wavelength Selective Cross-Connect Mesh Topology," *Proc. IEEE/OSA OFC'98*, Feb. 1998, pp. 61–62.

[185] Mahony, M. J., "Physical Limitations in Optical Cross-Connect Networks," *ECOC'95 Proc.*, Vol. 4, Tutorials, Brussels, Sept. 17, 1995, pp. 61–81.

[186] Cheng, Z., et al., "Crosstalk Limitation in Optical Cross-Connect Node with Filters," *J. of Opt. Com.*, Vol. 21, Oct. 2000, pp. 185–189.

[187] Pires, J. J., and L. G. C. Cancela, "Coherent Multipath Crosstalk in Benes and Dilated Benes Optical Switches," *LEOS'97 Proc.*, Vol. 2, San Francisco, CA, Nov. 10–13, 1997, pp. 544–545.

[188] Yu, C. X., W. -K. Wang, and S. D. Brorson, "System Degradation Due to Multipath Coherent Crosstalk in WDM Network Nodes," *IEEE J. Lightwave Technol.*, Vol. 16, No. 8, 1998, pp. 1380–1386.

[189] Shen, Y., K. Lu, and W. Gu, "Coherent and Incoherent Crosstalk in WDM Optical Networks," *IEEE J. Lightwave Technol.*, Vol. 17, No. 5, May 1999, pp. 759–764.

[190] Ho, K. -P., and J. Kahn, "Crosstalk Measurement and Reduction in Dense WDM Systems Using Subcarrier Tone Channel Identification and Linear Cancellation," *ICC'95 IEEE Proc.*, Seattle, WA, June 1995, pp. 287–291.

[191] Montgomery, J., "Optical Evolutions," *Telecommunications*, Feb. 2000, pp. 101–102.

[192] Koch, T. L., "WDM Sources and Receivers," *ECOC'95 Proc.*, Vol. 4, Tutorials, Brussels, Sept. 17, 1995, pp. 29–59.

[193] Langer, K. -D., et al., "Transparent Interconnection of Multi-Vendor Metropolitan Sub-Networks," *ECOC'2000 Proc.*, Vol. 4, Munich, Sept. 3–5, 2000, pp. 87–88.

[194] Eftimov, T., et al., "8-Channel Tunable Drop Device with Thermal Tuning for 100GHz Channel Spacing," *ECOC'98 Proc.*, Vol. 1, Madrid, Sept. 20–24, 1998, pp. 127–128.

[195] Bo, W., et al., "Fiber Gratings Based Optical Add/Drop Multiplexer with Low Interferometric Crosstalk," *International Conf. on Communication Tech. ICCT'98 Proc.*, Beijing, China, Oct. 22–24, 1998.

[196] Arecco, F., et al., "Acousto-Optic Devices in Add/Drop Multiplexer Nodes," *ECOC'98 Proc.*, Vol. 1, Madrid, Sept. 20–24, 1998, pp. 117–118.

[197] Barozzi, G., et al., "25 GHz AOTF Based Configurable Add/Drop Node," *ECOC'2000 Proc.*, Vol. 4, Munich, Sept. 3–5, 2000, pp. 57–58.

[198] Okamoto, K., "Fundamentals, Technology and Applications of AWGs," *ECOC'98 Proc.,* Vol. 2, Madrid, Sept. 20–24, 1998, pp. 9–47.

[199] Tachikawa, Y., et al., "Optical Add-Drop Multiplexers and Multi-Wavelength Optical Sources for WDM Transmission Systems," *NTT Review,* Vol. 10, No. 1, Jan. 1998, pp. 52–59.

[200] Oda, K., et al., "10 Channel × 10 Gbit/s over 500 km Optical FDM-Add/Drop Multiplexing Experiment Employing a 16-Channel Arrayed Waveguide-Grating ADM Filter," *ECOC'95 Proc.,* Vol. 1, Brussels, Sept. 17, 1995, pp. 59–62.

[201] Zhao, Y., et al., "IntraBand-Crosstalk Free Add/Drop Modules for Wavelength-Division-Multiplexed Ring Networks," *ECOC'2000 Proc.,* Vol. 4, Munich, Sept. 3–5, 2000, pp. 53–54.

[202] Ford, J., et al., "Wavelength-Selectable Add/Drop with Tilting Micromirrors," *LEOS'97 Proc.,* Post-deadline papers 2.3, San Francisco, CA, Nov. 10–13, 1997.

[203] Walker, J. A., "The Future of MEMS in Telecommunications Networks," *J. Micromech. Microeng.,* Vol. 10, IOP Pub. Ltd, U.K., 2000, R1–R7.

[204] Lewotsky, K., "Holographic Switch Tells Light Where to Go," *OE Magazine SPIE,* Jan. 2001, pp. 9–12.

[205] Mola, D. D., et al., "Tunable Add-Drop Filter in LPCVD Glass on Silicon Technology for AODM-Based Ring Networks," *ECOC'98 Proc.,* Vol. 1, Madrid, Sept. 20–24, 1998, pp. 123–124.

[206] Germann, R., "Tunable Optical Filter Adds and Drops at Will," *Fibre Systems Europe,* July 1999, p. 13.

[207] Horita, M., S. Tanaka, and Y. Matsushima, "Wavelength Tunable Optical Add and Drop Multiplexer Utilizing Coupled Semiconductor Waveguides," *ECOC'98 Proc.,* Vol. 1, Madrid, Sept. 20–24, 1998, pp. 115–116.

[208] Almström, E., S. N. Larsson, and H. Carldén, "Cascadability of Optical Add/Drop Multiplexers," *ECOC'98 Proc.,* Vol. 1, Madrid, Sept. 20–24, 1998, pp. 589–590.

[209] Evaldsson, P., et al., "METON Reports" *Workshop on WDM Network and Related Technologies Proc.,* 6 May 1998, pp. 40–45.

[210] Shimomura, H., and N. Henmi, "Wavelength Optical Add-Drop Multiplexer Configuration for 2.4 and 10 Gb/s Dual Bit Rates WDM Networks," *NFOEC'99, Tech. Proc.* Vol. 1, Chicago, Sept. 26–27, 1999, pp. 252–259.

[211] Diez, S., et al., "Novel Gain-Transparent SOA-Switch for High Bitrate OTDM Add/Drop Multiplexing," *ECOC'98 Proc.,* Vol. 1, Madrid, Sept. 20–24, 1998, pp. 461–462.

7

WDM Limits Caused by Optical Nonlinearities in Optical Fibers

7.1 Introduction

Considering that optical sources with wavelength specifications enabling smaller and smaller channel spacings are becoming available, and that components with almost arbitrarily high-channel counts are feasible, one could think that the number of channels would not be limited. Of course, this is not true: Optical nonlinearities set limits on the optical power that can be transmitted on the wavelength channels of the optical fiber.

The nonlinearities are related to:

- Index variation caused by intense electrical fields such as:
  - Self-phase modulation (SPM);
  - XPM;
  - FWM;
- Scattering effects such as:
  - Stimulated Brillouin scattering (SBS);
  - Stimulated Raman scattering (SRS).

All these nonlinear effects are locally rather small, but they become important by accumulation over long fiber lengths. They generally degrade the transmission quality through signal losses, intra- or interchannel crosstalks, pulse broadening or induced jitters. Of course the nonlinear effects can be advantageous in other cases such as: soliton transmission, amplification, and wavelength conversion. In this chapter, we consider the limiting effects only. The other effects are discussed in different chapters of this book (soliton: Section 2.9, amplification: Sections 5.2 to 5.9, conversion: Section 4.5).

7.2 SPM

In long-haul systems high powers are required to reach reasonable optical amplifier spacing at high bit rate. Index variations, due to the intense electrical field of the light signals, cause the Kerr effect to occur, and thus induces changes in the signal phase, leading to SPM. This index variation modulates the signal phase, which, in turn, modifies the signal spectrum. The pulse is broadened by SPM with a negative dispersion ("normal dispersion regime," Figure 7.1) and shortened by SPM with a positive dispersion ("anomalous dispersion regime"). In this later case, the temporal broadening due to chromatic dispersion is reduced by SPM. The phase change is about π rad for one watt of launched power after a propagation of 1 km on a typical classical SM fiber at 1.55 μm [1].

Of course, SPM raises problems in monochannel transmission as well as in multiplexed transmission, except in soliton transmissions and in a few other specific cases to the contrary, in which this effect is used (for instance the Kerr nonlinearity in normal dispersion fibers can be used for improving the phase margin in optical RZ receivers [2]).

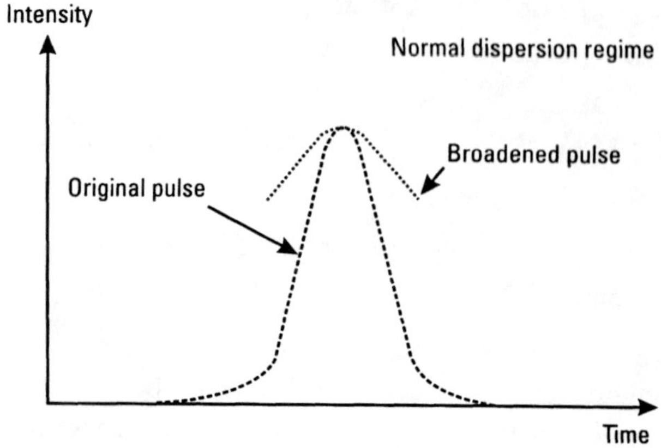

Figure 7.1 The pulse is broadened by SPM in a negative dispersion fiber.

The electrical field (E) intensity-dependent refractive index is described as:

$$n = n_0 + n_2 |E|^2$$

where n is the perturbed index in the fiber core, n_0 is the unperturbed index, n_2 is the nonlinear-field-dependent refractive index coefficient. The refractive index change induces a time phase-shift across the pulse as it propagates along the fiber:

$$\Delta\phi = \Delta n \mathrm{L}$$

The instantaneous angular frequency, ω is changed according to:

$$\partial\omega = \partial\Delta\phi / \partial t$$

For Gaussian pulses with initial width $\Delta\omega$ the width broadening is:

$$\partial\omega \approx 0.86\ \Delta\omega\Delta\phi_{\max(\text{Radians})}$$

Due to this effect, the rising edge of a Gaussian pulse is "red"shifted while the trailing edge is "blue" shifted. This frequency chirp can be used in negative group-velocity dispersion fibers in which it can compensate the effect of dispersion ("blue" traveling faster than "red") under certain conditions (see Section 2.9). However, most often, SPM will set limits on the transmission quality. Nonlinear coefficient round-robin measurements were performed in Japan and in the United Kingdom for various nonzero dispersion-shifted fibers (NZDSF) and large effective area (A_{eff}) dispersion-shifted fibers (LEDSF) [3]. The averaged values of n_2/A_{eff} at random polarization states were measured to be ≈ -4.7, ≈ -4.0, and $\approx -3.0 \times 10^{10}$ W^{-1} for DSF, NZDSF, and LEDSF, respectively. The coefficient n_2 was the same for each fiber type: $n_2 = -2.20 \times 10^{-20}$ m^2 W^{-1}. Measurements on Corning SMF 28 fibers over the wavelength range 1,534 to 1,560 nm [4] give an average value of: $n_2 = -(2.53 \pm 0.09) \times 10^{-20}$ m^2 W^{-1}. Other information on nonlinear coefficients is given in [5–8].

The nonlinear coefficient is $\gamma = 2\pi n_2 / (\lambda A_{\text{eff}})$. This coefficient decreases when the effective fiber core increases. The nonlinear threshold is about 1.1 W.km for G.653 and G.655 fibers and 1.6 W.km for G.652 fibers.

Note that the effective area A_{eff} of an SM optical fiber is defined in ITU standards [9, 10] as:

$$A_{eff} = \frac{2\pi\left[\int_0^\infty |E_a(r)|^2 \, rdr\right]^2}{\left[\int_0^\infty |E_a(r)|^4 \, rdr\right]}$$

where $E_a(r)$ is the mode-field distribution on a radius r.

Typical values of A_{eff} at 1.5 μm are 80, 54, 83, 46, and 68 μm^2 for attenuation flattened fibers, dispersion flattened fibers, matched cladding fibers, dispersion-shifted fibers, and LEAF fibers, respectively.

In today's high-transmission distance links, the effects of dispersion and self-phase modulation have to be considered together for system optimization by proper dispersion management [11].

It was shown that duobinary modulations offer large dispersion tolerance, increase the bandwidth efficiency, and eliminate SBS effects (see Section 7.5) [12–15]. However, it was also demonstrated, in a comparison of the SPM limits of duobinary and binary codings in dispersion-managed and uncompensated systems, that "beyond a critical power level the advantage of the duobinary format with respect to dispersion tolerance was substantially reduced" [16].

SPM limits the maximum link length at high bit rates. For a given bandwidth and distance, the optical signal-to-noise ratio can always be increased with more launch power but only up to the point where the fiber nonlinearity begins to penalize the signal. The transmission limits imposed by SPM on single-channel multispan transmission links for return-to-zero (RZ), and nonreturn-to-zero (NRZ) data format from 10 to 80 Gbps were investigated by G. Mohs, et al. [17].

RZ seems advantageous for systems that are limited by single-channel effects such as SPM. However, considering other nonlinear interchannel crosstalks occurring in WDM systems (namely, XPM, FWM, and SRS: see Sections 7.3, 7.4, and 7.6), NRZ performs equally or better than RZ coding [18].

Remark: SPM is not always a problem. We have already seen that SPM could be very useful in some cases such as soliton propagation. It is also worth noting that SPM in dispersion-shifted fibers can be used for the regeneration of high bit-rates signals [19]. It is possible to suppress the noise in the "zeros" and the amplitude fluctuation in the "ones" of RZ optical TDM data streams using SPM.

Using this effect, simultaneous time demultiplexing and regeneration was achieved from 40 Gbs OTDM signals [20]. Also, SPM in a SOA was used for pulse extinction ratio improvement in 40 Gbps OTDM applications [21].

7.3 XPM

7.3.1 Interchannel XPM

Interchannel XPM corresponds to the phase variation arising from the index variation induced by one signal at wavelength λi in another codirectionnal channel at wavelength λj. This effect sets a limit to the peak power that can be coupled to the fiber. For instance, a set of 2.5-Gbps channels around 1,550 nm, with a spacing of about 0.5 nm, cannot be used with more than 10 mW per channel, on a standard SMF fiber (without dispersion shift) on a long-distance (720 km) good-quality link. The XPM interference is almost proportional to $1/\Delta\lambda$ as calculated and experimentally demonstrated by Yu, Jeppesen, and others. ($\Delta\lambda$ is the spacing between channels). This was verified numerically with bit rates as high as 40 Gbps. They also demonstrate that the difference in crosstalk between NRZ and RZ modulation format with a duty cycle larger than a given value is quite small [22]. These conclusions were verified for standard SMF fibers (dispersion 17 ps/nm/km at 1,560 nm) and TrueWave fibers (dispersion 5 ps/nm/km at 1,560 nm).

In point-to-point links, the effect of XPM can be minimized by controlling the phases between the clock signals of the different channels [23]. Quite fortunately, XPM saturates with increasing channel numbers due to velocity mismatch along the fiber [24]. R. A. Saunders, et al. have shown how to compensate in part the XPM in a 10 Gbps system over nonzero dispersion-shifted fibers using a phase prechirp on the interfering channel [25].

7.3.2 Intrachannel Cross-Phase Modulation

In addition to XPM between channels, intrachannel cross-phase modulation (IXPM) has been shown to lead to pulse timing jitter at high bit rates. The intrachannel nonlinear effect in 40-Gbps WDM transmission over standard-fiber links can be potentially more harmful than XPM between channels [26]. However, the problem can be limited by a precompensation with dispersion D_{PRE} (ps/nm). Killey, et al. proposed a simple analytical rule to calculate the optimum dispersion map for practical systems [26]:

$$D_{PRE} = \frac{-D_{SMF}}{\alpha} \ln\left[\frac{2}{1+e^{-\alpha L}}\right]$$

Where α is the fiber loss, D_{SMF} is the fiber dispersion, and L is the span length. In their practical system they showed that both IXPM and IFWM were minimized with $D \approx -200$ ps/nm.

It was shown that all Kerr nonlinearities in long fiber spans could be canceled using a $LiNbO_3$ phase conjugator and Raman amplification [27]. Ultra-low duty-cycle RZ transmission has a very-high tolerance to the fiber nonlinearities [28, 29]. Clausen, et al. demonstrated that the penalties reduce with the decrease of the pulse width and with the increase of the fiber dispersion and that a proper predispersion combined with counter-propagating Raman amplification can reduce significantly the nonlinear impairments [30].

In dense dispersion-management soliton, WDM transmission periodic spans of an SM fiber followed by reverse-dispersion fiber, plus a single span of an SM fiber to adjust the path-averaged dispersion can be used. The span-averaged dispersion can be tuned in each span to control the interchannel XPM, while intrachannel interactions may be reduced by polarization multiplexing [31].

7.3.3 Nonlinear Channel Depolarization Through XPM

A nonlinear channel depolarization through XPM was first observed in very-long-haul (a few thousand kms) soliton transmission at 10 Gbps [32]. In general, the linear polarization along the transmission line induced by PMD and PLD can be compensated to some extent as they produce relatively slow or static changes. In contrast, the timescale of nonlinear polarization changes is very short (bit-period) scale. It cannot be easily compensated for. If polarization-sensitive receivers are used, the polarization variations are converted into amplitude fluctuations reducing the performances of the system. In polarization-state multiplexing in which orthogonally polarized adjacent channels are used in order to push the spectral efficiency of DWDM, polarization-sensitive receivers are used. In such systems, the nonlinear depolarization through XPM set limits to the performances. For example, B. C. Collings and L. Boivin observed a substantial reduction in the power margin by this effect, in a five-channel spaced by 50 GHz, 10 Gbps, NRZ, only 360-km system using orthogonally launched adjacent signals (at about 4 dBm per channel) and polarization-sensitive receivers [33].

7.4 FWM

7.4.1 Interchannel FWM

Three signal frequencies ν_i, ν_j, ν_k, closely spaced can generate a mixing product at a fourth frequency $\nu_i + \nu_j - \nu_k$. This effect is called four-wave mixing. In principle, FWM may be caused by the mixing of only two frequencies. However, two-frequency-mixing products have a weaker intensity than three-frequency-mixing products. FWM corresponds to beatings between signals at different frequencies. Two signals at frequencies ν_i and ν_j generate harmonics at frequencies $\nu_i - (\nu_j - \nu_i)$ and $\nu_j + (\nu_j - \nu_i)$, (Figure 7.2). These harmonics lead to additional noise that was particularly harmful if they landed on other adjacent WDM channels. From N original wavelengths $N^2(N-1)/2$ new wavelength are created.

If $\nu_i - \nu_j = \Delta\nu$, the first harmonics are at frequencies $\nu - \Delta\nu$ and $\nu + \Delta\nu$; consequently, the crosstalk is maximum when WDM channels are equally spaced in frequencies (Figure 7.3).

Unfortunately, the ITU standard sets a grid of frequencies with a constant $\Delta\nu$ spacing (100 GHz). But it is still possible to use unequally spaced channels selected on the ITU grid to reduce the FWM crosstalk. It was shown that a slight detuning of about 5 GHz from the center frequency of a given channel in a 100-GHz-spacing 16-channel multiplexing can decrease the bit error rate (BER) significantly. Such a detuning is compatible with the ITU recommendations on frequency tolerance over the frequency grid [34].

The FWM efficiency is proportional to $P_{ch}/(\Delta\lambda^4 D^2)$. (Where P_{ch} is the power in the interfering channels, $\Delta\lambda$ is the spacing and D is the dispersion). The power generated in the adjacent channels is also proportional to

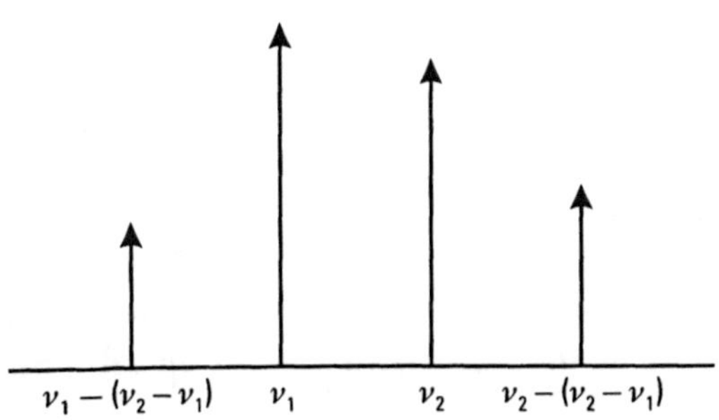

Figure 7.2 Signals at frequencies ν_1, ν_2 and their FWM harmonics.

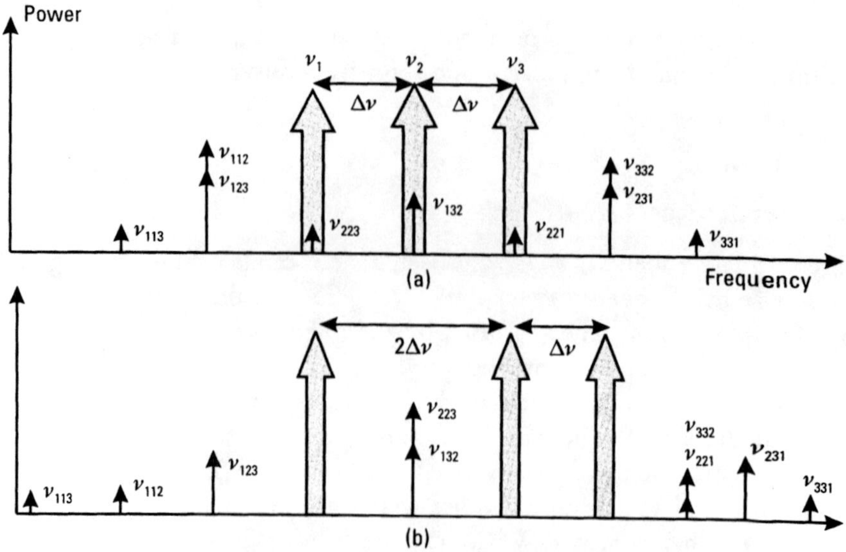

Figure 7.3 Signals at frequencies ν_1, ν_2, ν_2 and their 9 FWM harmonics, (a) with a constant spacing three FWM products generated fall on signal channels, and (b) with frequency spacing $\Delta\nu$ and $2\Delta\nu$ no products generated fall on signal channels.

$(1/A_{eff})^2$. The FWM is an issue for G.653 fibers but less of an issue for G.655 with larger A_{eff}. The FWM is generally not an issue for G.655 fibers.

- FWM increases with the optical signal power so, in practice, the maximum distance usable between repeater amplifiers is reduced by FWM.
- FWM increases when the frequency spacing decreases so it sets practical limits to this spacing. FWM corresponds to real limitations in coherent multiplexing. For instance, Waart and Braun [35], tackling the FWM problem, demonstrated that, with a multiplexed link at 100 channels, 5-GHz-spaced, where it is intended to limit the crosstalk at −20 dB, the input power must be limited from 0 to −5 dBm for 5 km, and to −10 dBm for 100 km. D. A. Cleland, et al. showed that FWM is the dominant consideration for the determination of channel spacing in long-distance links (typically 560 km, 2.4 Gbps) [36].
- FWM depends on the relative polarization of the beating signals.

- The effect of fiber-chromatic dispersion on FWM is critical [37] and a negligible penalty with conventional step index fiber may become high with dispersion-shifted fibers [38].

Actually, FWM is the main limitation in DWDM systems using dispersion-shifted fibers near the zero-dispersion wavelength region, because in such a case the harmonics remain in phase over long distances with the original signals: The effect is cumulative. With standard fibers used on most networks (dispersion 16 ps/nm/km) the FWM effect is much less important.

In a bidirectional zero-dispersion fiber, the number of channels that can be used in a given spectral region with low-enough nonlinearity can be increased using sets of frequencies, different in both directions, with unequal spacing, on the constant-spacing ITU grid.

For example, T. Sato and T. Horiguchi transmitted good quality signals at 10 GB/s over 2*70 km using five frequencies unequally spaced in both directions on a grid of 11 equally spaced frequencies with 200-GHz spacing [39].

Fibers with small but nonzero chromatic dispersion can be used in order to reduce the nonlinear effects. For instance, commercially available fibers show the following median values (in C band):

- TrueWave Fibers (Corning):
  - Attenuation 0.20 dB/km;
  - Chromatics dispersion 2.4 ps/nm-km;
  - Polarization-mode dispersion 0.05 ps/km;
  - Nonlinear effective area 54 μm^2;
  - Core/clad concentricity 0.2 μm.

- TrueWave Fibers-RS (Corning):
  - Chromatics dispersion 3.7 ps/nm-km;
  - Slope 0.058 ps/nm^{2}km;
  - Nonlinear effective area 65 μm^2.

- E-LEAF (Corning):
  - Chromatics dispersion 4.2 ps/nm-km;
  - Slope 0.084 ps/nm^{2}km;
  - Nonlinear effective area 72 μm^2.

- Teralight fibers (Alcatel):
    - Chromatics dispersion 8.0 ps/nm-km;
    - Slope 0.058 ps/nm^{2}km;
    - Nonlinear Effective Area 65 μm^2.

(*Note:* Other fibers with reduced nonlinear effects are available.)

For example, in 1995 Chraplyvy and Tkach demonstrated that four channels spaced at 2, 1, and 1.5 nm near 1,546 nm with a launch power of 3 dBm per channel gave redhibitory FWM after 25 km of dispersion-shifted fiber but did not showed noticeable FWM after 50 km of TW. It was reported that TW's dispersion is large enough to suppress FWM, but small enough to allow 8 wavelengths, each at 2.5 Gbps to be transmitted 1,000 km without dispersion.

Conversely, since 1998, on fibers with zero dispersion at 1,550 nm already installed, FWM had been reduced using sources with longer wavelengths well adapted to new amplifiers with gain in the 1,570- to 1,600-nm region. For a standard 1,550-nm zero-dispersion fiber the group-velocity dispersion is 2.0 to 2.9 ps/nm/km in the 1,570- to 1,600-nm region.

For instance, M. Jinno, et al. were able to design a high-quality multiplexed link with 8 channels equally spaced at 200 GHz, each at 10 Gbps bit rate, without repeater over 160 km [40].

In a very significant experiment, A. K. Srivastava, et al. [41] were able to transmit 64 channels, at 10 Gbps each, 50 GHz spaced over 500 km in the 1,570- to 1,610-nm band (L band of Erbium-doped amplifiers) with bit error rates smaller than 10^{-9}. Moreover the XPM, normally significant, remains small enough in this demonstration.

7.4.2 Intrachannel FWM

In very-high bit-rate dispersion-managed transmissions IXFWM is one of the limiting factors [26, 42]. It leads to overlapping of the pulses broadened by the fiber dispersion generating "shadow" or ghost pulses in the zeros. As IXPM, IXFWM can be reduced with the design of an optimum-dispersion map. Normally the same value of dispersion precompensation reduces these two distortions.

In [43], rules are derived for an optimized design of WDM links at 10 Gbps per channel, considering XPM and SRS. The optimum design will take into account the fact that XPM saturates due to group-velocity mismatch, but that conversely SRS increases, when the number of channels is increased.

7.5 SBS

SBS corresponds to the interaction of the optical waves with the acoustic waves that occur in the fiber. A contradirectional scattering at frequencies smaller than the incident frequency takes place. The threshold of this nonlinear effect depends on the source spectral width and on the power density in the fiber. The launch power limitation due to SBS does not depend on the number of channels and is more than a few dBm in typical applications [44].

The SBS threshold power as given by Y. Aoki, et al. [45] is:

$$P_{thr} = \frac{21 A_{eff}}{g_B L_{eff}} \left[1 + \frac{\Delta\nu_P}{\Delta\nu_B} \right]$$

where $\Delta\nu_p$ and $\Delta\nu_B$ are the line-widths of the input laser and of the Brillouin line, A_{eff} and L_{eff} are the effective core area and the effective length of the fiber, and g_B is the Brillouin gain in the fiber.

In G.652 fibers the shift is approximately 11 GHz (+0.088 nm) and $\Delta\nu_B \approx$ 30MHz. The gain coefficient is about 4 10^{-9} cm/W. In G.653 fibers the shift is approximately 10.7 GHz (+0.085 nm) and $\Delta\nu_B \approx$ 80MHz. The gain coefficient is about 3 10^{-9} cm/W [46].

In the large effective core area fibers in which the power density is smaller that in standard fibers, the threshold is larger and the SBS effect is smaller. The typical SBS threshold is about 10 dBm for a CW source. The threshold increases by 3 dB when the source is externally modulated.

It is also often possible to broaden the laser line-width by direct or external modulation when necessary: In that way, the SBS threshold can generally be increased by several decibels. The SBS threshold can also be increased by 3 dB by a polarization-multiplexing technique [47].

A duobinary modulation has proven to be a simple and efficient way to increase the SBS threshold [48].

As its threshold can be increased by different ways, generally SBS is not a big problem in wavelength-division multiplexed transmission links, and the impact on the system performance is relatively small.

7.6 SRS

Raman scattering is related to the absorption of energy from a photon followed by loss of a part of its energy to a mechanical vibrational state in the medium and a reradiation of light at longer wavelength (lower energy photon) (Figure 7.4). The process can be stimulated when another photon at this longer wavelength is incident on the excited molecule. From this incident photon a second photon will be emitted. This process corresponds to the stimulated Raman amplification, used in some lightwave amplifiers.

The problem of limitation of launch power by SBS is being solved by different means such as line broadening and/or by other means, SRS sets another limit [49–54].

In Raman scattering, the fiber, excited at a given wavelength, reemits a scattering spectrum at the same wavelength (Rayleigh scattering), which is not a problem, but also at other wavelengths, longer or shorter than the excitation wavelength, which are termed, respectively, Stokes and anti-Stokes Raman scattering. The Stokes spectrum only has enough intensity to become a constraint with cumulative effects along the fiber. The Raman peak is about at a 13.2-THz distance from the excitation line (Figure 7.5). SRS has two negative impacts: It induces crosstalk between channels by power exchange, and it limits the maximum launched power in WDM channels through power depletion. This happens at power above the SRS threshold (≈ 1W).

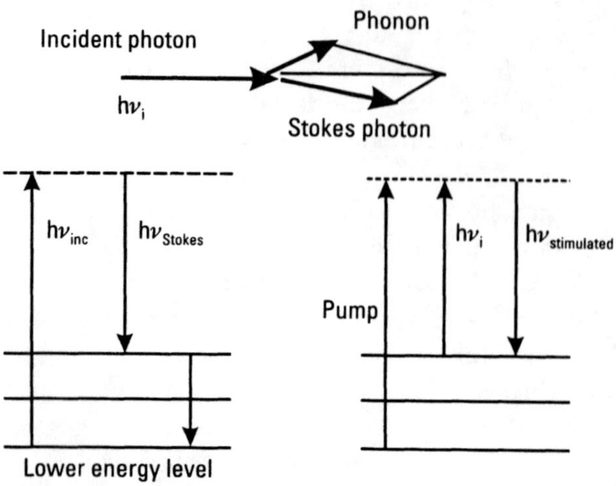

Figure 7.4 Stokes Raman scattering and stimulated Raman scattering energy levels.

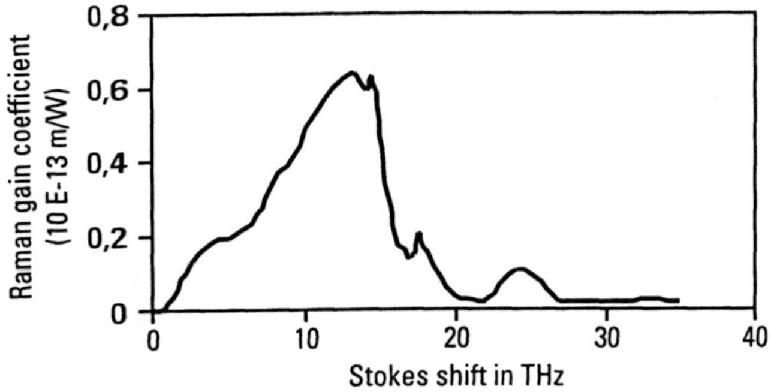

Figure 7.5 Raman gain coefficient of silica glasses in the 1,500-nm region.

The Raman gain coefficient and the Raman threshold were measured for different practical transmission fibers [55, 56]: standard single-mode fiber (SSMF), pure silica-core fiber (PSCF), LS fiber, and TrueWave fiber (TW), as shown in Table 7.1.

The combination of the different paramcters gives different thresholds from 30.2 dBm on the standard fiber down to 26.6 dBm on the TW.

Raman crosstalk may not result in practical consequences in some cases. P. Niay, et al. [57], working at about 1.3 μm, with two 430 cm^{-1}-spaced channels (not far from the worst case), for −3 dBm injected powers (in such a case, stimulated Brillouin and Raman effects are negligible) and a 100% analogic modulation rate, showed that the crosstalk is about -30 dB for 15 km of SM fiber. It would reach −25 dB for an analogous system at 1.5 μm on 25 km in the same conditions. However, the SRS crosstalk can become a problem at total input powers of about 100 mW and above. For example, on 32-channels transmission with 100-GHz spacing in C-band DWDM over 100 km, the transmitted power penalty on the lower wavelength channel is 0.7 and 2.3 dB with −10 and 5.6 dBm per channel, respectively [58].

Table 7.1
Raman Gain Coefficients, Losses, and Mode Areas of a Few Fibers

Fiber Type	**SSMF**	**PSCF**	**LS**	**TW**
Mode area μm^2/loss dB/km	78/0.21	75/0.17	53/0.22	53/0.22
Raman coef. 10^{-14} m/W	5.3	6.0	6.5	6.0

S. Chi and S. C. Wang [59] calculated the limitation in bit rate × maximum length of high-density, optically multiplexed links related to Raman scattering on fibers with dispersion shifted at 1.55 μm only. With 10^{-13} Joule input pulses, they found a bit rate × maximum length of 3.27 10^5 km Gbps. This best case could be obtained with 0.2-nm-channel spacing and 240 wavelengths. See also [60].

In SRS [61, 62], the high-frequency channels act as pumps for smaller frequency channels. The effect depends on the distance between channels. The effect is maximized with a frequency distance of ≈ 13.2 THz (≈ 100 nm in the 1.5 μm wavelength region) between the sources. This corresponds to the distance between the Rayleigh line and the main Raman peaks of the silica material. As the distances in DWDM are much smaller (0.4 to 10 nm), SBS has, in general, a very little impact on the signal quality [63].

However, in case of many-channel WDM, the total power and the spectral range become large enough to induce significant SRS. The SRS coupling can be calculated from the following equation [64]:

$$\frac{dI_j}{dz} = \sum_i g_R(\nu_i - \nu_j) I_i I_j - \sum_k \frac{\nu_j}{\nu_k} g_R(\nu_j - \nu_k) I_j I_k - \alpha_j I_j$$

Where $\nu_i > \nu_j > \nu_k$, I_j is the intensity at frequency ν_j, g_R $(\nu_i - \nu_j)$ is the Raman gain profile in Figure 7.5. The first sum is the gain of beam j from beams at higher frequencies. The second sum is the loss to beams at lower frequencies. The third term is the attenuation along the fiber. Using this equation McIntosh, et al. observed "significant channel depletion due to the SRS noise above 1W of total input power." However, they also showed that "inclusion of a high-frequency pass filter greatly increases the SRS threshold of the Raman noise that causes the channel depletion."

Of course, in a very wide WDM window (100 nm and above), the SRS needs to be compensated. With the new gain-shifted thulium-doped fiber amplifier, the transmission window can be extended towards lower wavelengths of 1.48 μm. Using, simultaneously, S band (1,478 to 1,512 nm), C band (1,531 to1,563 nm), and L band (1,568 to 1,605 nm) becomes feasible. One hundred thirty-nm-wide WDM transmission over 125 km of SMF fiber with 400-GHz-spaced 10 Gbps signals was demonstrated [65]. In this experiment, the SRS needed to be compensated mainly in the S band (Raman scattering is more important at shorter wavelengths). This was done with a distributed Raman amplification with a counter-propagating pump at 1.4 μm in a thulium-doped fiber.

In [66] Keang-PoHo evaluates both power penalty and power limit induced by Raman crosstalk in single- and multispan WDM systems.

7.7 Conclusions

In optical-fiber transmission systems an increase in capacity is obtained via an increase of the channel data rate and the number of wavelength channels. Both correspond, necessarily, to an increase in the required launch power. However, this power is limited by nonlinearities.

Today in very-high data rate 40 Gbps links, on standard SM fibers, the main theoretical limitation is SPM. SPM is a major problem on monochannel as well as on WDM optical transmission. (However, in most cases the polarization-mode dispersion of practical fibers limits the data rate to lower values.)

In DWDM systems operating at smaller data rates, the main penalty arising from nonlinerarities is related to FWM in fibers when used near zero of dispersion. This limits the launch power. FWM has a smaller impact on standard nondispersion-shifted SM fiber. But FWM becomes a problem with dispersion-compensated fibers. XPM, SBS, and SRS need also to be taken into account. As XPM saturates with an increasing number of channels, SRS increases with wider wavelength range. SBS has a small impact as the line-width of the sources that can generally be broadened in order to reduce this particular nonlinear effect.

In order to reduce the launched power, the use of a contrapropagating pump-distributed Raman gain optical amplification is recommended (Figure 7.6). In such a case it can be necessary to reduce the spontaneous Raman scattering in the pump laser itself by insertion of filters on the line [67].

Fibers with larger effective core areas allow lower power density and thus lower nonlinear effects for a given transmitted power. However, the chromatic dispersion of the line needs to be managed: hybrid fiber spans with a large area, non-zero dispersion fiber, and a dispersion-compensating fiber is often the best choice for high data-rate WDM-TDM links (Figure 7.7).

In a given system architecture, the best solution will correspond to a trade-off between the dispersion compensation of the fibers, the signal launch powers, and the cumulative nonlinear crosstalks [68–76]. Each month, better multi-Tbps long-distance DWDM links, with more wavelengths, higher bit rates, longer spans, and in which the nonlinearities

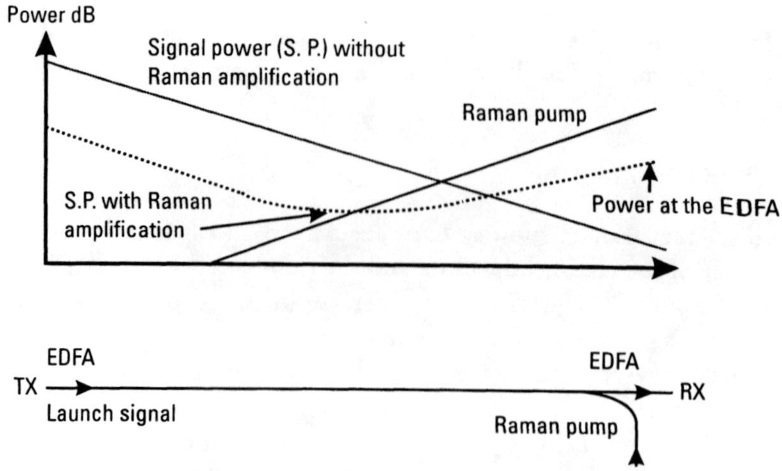

Figure 7.6 Lower launch power from the transmitter TX, reduces nonlinear effects. However, it is necessary to get enough power at the input of the optical amplifier to reduce noise effects on the receiver RX. Adding distributed gain from a contrapropagating Raman pump in EDFA amplifier links allows much longer DWDM transmission spans (typically three times longer).

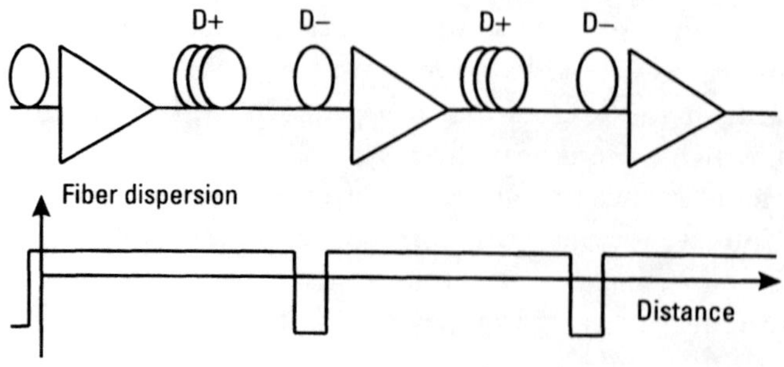

Figure 7.7 Dispersion management. This is a powerful technique for nonlinearity effect limitation. Long spans of standard fibers or fibers with a low average dispersion (NZDSF), and segments of fibers with a high local dispersion of opposite sign, give only weak signal distortions while reducing the nonlinear impairments.

problems find adequate solutions, are tested. Some examples given during the post-deadline session of the European Conference on Optical Communication in September 2000 give an idea of the performances at this time [77–84].

7.8 Additional Note

See other useful references on nonlinear effects not directly cited above:

- XPM [85];
- FWM, unequal spacing [86–91];
- FWM, effect of dispersion [92];
- FWM, intrachannel effects [93];
- SPM and XPM [94].

References

[1] Kashyap, R., "Nonlinear Optical Fibers," *Trends in Optical Fiber Metrology and Standards*, D. D. Soares (ed.), Kluwer Academic Publishers, 1995.

[2] Suzuki, et al., "Improvement of a Phase Margin in Optical RZ Receiver Using Kerr Nonlinearity in Normal Dispersion Fibers: Experimental Verification," *ECOC'2000 Proc.*, Vol. 4, Munich, 2000, pp. 49–50.

[3] Namihira, Y., "Nonlinear Coefficient Round Robin Measurements for Various Dispersion Shifted Fibers in Japan and U.K.," *ECOC'2000 Proc.*, Vol. 3, Munich, 2000, pp. 97–98.

[4] Dudley, J. M., et al., "Simultaneous Measurement of the Dispersion and Nonlinearity of Standard Fibre Using Frequency Resolved Optical Gating," *ECOC'97, IEE Proc.*, Vol. 3, Edinburg, 1997, pp. 307–310.

[5] Kim, K. S., et al., "Measurement of the Non-Linear Index of Silica Core and Dispersion-Shifted Fibers," *Optics Lett.*, Vol. 19, No. 14, 1994, pp. 257–259.

[6] Namihira, Y., "Highly Accurate Nonlinear Coefficient Measurements by SPM Method for DSFs and Large Effective Area Fibers at 1.55 μm," *SOFM'98 Proc.*, NIST, 1998, pp. 83–86.

[7] Boskovic, et al., "Direct Continuous-Wave Measurement of n_2 in Various Types of Telecommunication Fiber at 1.55μm," *Optics Lett.*, Vol. 21, No. 24, 1996, pp. 1966–1968

[8] Wada, A., "Measurements of Nonlinear-Index Coefficients of Optical Fibers Through the Cross-Phase Modulation Using Delayed Self-Heterodyne Technique," *ECOC'92 Proc.*, Paper, Mo B1.2, 1992, pp. 45–48.

[9] *ITU COM,* "15-273-E Definition and Test Methods for the Relevant Parameters of Single-Mode Fibres," Appendix on Nonlinearities for G.650, (1996).

[10] Billington, R. S., et al., "Effective Area of Single-Mode Optical Fibres: A Comparison of Four Measurement Techniques," *ECOC'2000 Proc.*, Vol. 3, Munich, 2000, pp. 139–141.

[11] Färbert, A., et al., "Optimal Dispersion Management for Long-Haul Optical Transmission Systems," *ECOC'99 Proc.*, Nice, France, 1999, pp. 154–155.

[12] Yonenaga, K., and S. Kuwano, "Dispersion-Tolerant Optical Transmission System Using Duobinary Transmitter and Binary Receiver," *IEEE J. Lightwave Technol.*, Vol. 15, No. 8, 1997, pp. 1530–1537.

[13] Penninckx, D., "Dispersion-Tolerant Modulation Techniques for Optical Communications," *ECOC'98 Proc.*, Vol. 1, Madrid, 1998, pp. 509–510.

[14] Wichers, M., and W. Rosenkranz, "Optical Duobinary Modulation Schemes Using a Mach-Zender Transmitter for LightWave Systems," *ICTON'99 Proc.*, Kielce, Poland, Paper We.B.1, 1999.

[15] Walklin, S., and J. Conradi, "On the Relationship Between Chromatic Dispersion and Transmitter Filter Response in Duobinary Optical Communication Systems," *IEEE Photon. Technol. Lett.*, Vol. 9, No. 7, 1997, pp. 1005–1007.

[16] Kaiser, W., et al., "SPM Limit of Duobinary Transmission," *ECOC'2000 Proc.*, Vol. 3, Munich, 2000, pp. 31–32.

[17] Mohs, G., "Maximum Link Length Versus Data Rate for SPM Limited Transmission Systems," *ECOC'2000 Proc.*, Vol. 3, Munich, 2000, pp. 191–192.

[18] Fürst, C., et al., "Comparison of Nonlinear Crosstalk for RZ and NRZ Coding in WDM Systems," *ECOC'2000 Proc.*, Vol. 3, Munich, 2000, pp. 193–194.

[19] Mamyshev, P. V., "All-Optical Data Regeneration Based on Self-Phase Modulation Effect," *ECOC'1998 Proc.*, Vol. 1, Madrid, 1998, pp. 475–476.

[20] Yu, J., and P. Jeppesen, "Simultaneous All-Optical Demultiplexing and Regeneration of a Channel from a 40 Gb/s OTDM Signal Based on SPM and XPM in Dispersion Shifted Fibers," *ECOC'2000 Proc.*, Vol. 1, Munich, 2000, pp. 21–22.

[21] Nielsen, M. L., et al., "Pulse Extinction Ratio Improvement Using SPM in a SOA for OTDM Systems Applications," *ECOC'2000 Proc.*, Vol. 1, Munich, 2000, pp. 23–24.

[22] Yu, J., and P. Jeppesen, "Investigation of Cross-Phase Modulation in WDM Systems with NRZ and RZ Modulation Format," *ECOC'2000 Proc.*, Vol. 3, Munich, 2000, pp. 199–200.

[23] Cartaxo, A., and M. O'Mahony, "Impact of Cross-Phase Modulation on WDM System Performance Taking the Relative Delay Between Channels into Account," *NOC'98 Proc.*, Amsterdam, The Netherlands, IOS Press, June 1998, pp. 259–266.

[24] Hanik, N., et al., "Optimized Design of Transparent Optical Domains," *ECOC'2000 Proc.*, Vol. 3, Munich, 2000, pp. 195–196.

[25] Saunders, R. A., et al., "Method of Compensating Cross Phase Modulation in 10Gb/s WDM Systems over Nonzero Dispersion Shifted Fiber," *Electron. Lett.*, Vol. 34, No. 18, September 3, 1998, pp. 1767–1769.

[26] Killey, R. I., H. J. Thiele, and P. Bayvel, "Improving the Performance of 40 Gb/s-Based WDM Transmission over Standard Fibre," *ECOC'2000 Proc.*, Vol. 3, Munich, 2000, pp. 29–30.

[27] Brener, I., et al., "Cancellation of All Kerr Nonlinearities in Long Fiber Spans Using $LiNbO_3$ Phase Conjugator and Raman Amplification," *OFC 2000 Proc.*, Baltimore, MD, Post-deadline paper PD33.

[28] Gnauck, A. H., et al., "Highly Dispersive Pulses for 40 Gbit/s Transmission over Multiple 120 Km Spans of Conventional Single-Mode Fiber," *Proc. LEOS Annual Meeting*, Post-deadline paper PD1.2, San Francisco, CA, 1999.

[29] Mikkelsen, B., et al., "Unrepeatered Transmission over 150 Km of Non-Zero-Dispersion Fiber at 100 Gbit/s with Semiconductor Based Pulse Source, Demultiplexer and Clock Recovery," *Electron. Lett.*, Vol. 35, 1999, p. 1866.

[30] Clausen, C. B., A. Mecozzi, and M. Shtaif, "Nonlinear Intra-Channel Effects: System Impairments and Their Remedy," *ECOC'20 00 Proc.*, Vol. 3, Munich, 2000, pp. 33–34.

[31] Wabnitz, S., and F. Neddam, "Control of Cross-Phase Modulation in Dispersion Flattened Soliton Systems," *ECOC'2000 Proc.*, Vol. 3, Munich, 2000, pp. 37–38.

[32] Mollenauer, L. F., J. P. Gordon, and F. Heismann, "Polarization Scattering by Soliton-Soliton Collision," *Opt. Lett.*, Vol. 20, 1995, pp. 2060–2062.

[33] Collings, B. C., and L. Boivin, "Nonlinear Polarization Evolution on the Time Scale of a Single Bit Period in a Fiber Optic Transmission Line," *ECOC'2000 Proc.*, Vol. 3, Munich, 2000, pp. 43–44.

[34] Boskovic, A., S. Ten, and V. L. da Silva, "FWM Penalty Reduction in Dense WDM Systems Through Channel Detuning," *ECOC'98 Proc.*, Vol. 1, Madrid, 1998, pp. 163–164.

[35] Waarts, R. G., and. R. P. Braun, "System Limitations Due to Four-Wave Mixing in Single Mode Optical Fibres," *Electron. Lett.*, Vol. 22, No. 16, July 31, 1986, pp. 873–875.

[36] Cleland, D. A., et al., "Limitations of WDM Transmission Over 560 Km Due to Degenerate Four Wave Mixing," *Electron. Lett. (U.K.)*, Vol. 28, No. 3, 1992, pp. 307–309.

[37] Walker, G. R., et al., "Effect of Launch Power and Polarization on Four-Wave Mixing in Multichannel Coherent Optical Transmission System," *Electron. Lett. (U.K.)*, Vol. 28, No. 9, April 23, 1992, pp. 878–879.

[38] Walker, G. R., et al., "Effect of Fiber Dispersion," *Electron. Lett. (U.K.)*, Vol. 28, No. 11, 1992, pp. 989–999.

[39] Sato, T., and T. Horiguchi, "A Novel Dense-WDM Optically Repeated Transmission System in the Zero-Dispersion Region Employing Bi-Directional USCA, FWM Induced Conjugate Light Filtering and a Bi-Directional Optical Amplifier," *ECOC'98 Proc.*, Vol. 3, Madrid, Sept. 20–24, 1998, pp. 151–152.

[40] Jinno, M., et al., "WDM Transmission Technologies for Dispersion-Shifted Fibers," *IEICE Trans. Electron.*, Vol. E81-C, No. 8, August 1998, pp. 1264–1274.

[41] Srivastava, A. K., "L-Band 64*10 Gb/s DWDM Transmission over 500 Km DSF with 50 GHz Channel Spacing," *ECOC'98 Proc.*, Vol. 3. Madrid, Sept. 20–24, 1998, pp. 73–75.

[42] Merlaud, F., and S. K. Turitsyn, "Intra-Channel Four Wave Mixing and Ghost Pulses Generation: Time Domain Approach," *ECOC'2000 Proc.*, Vol. 3, Munich, 2000, pp. 35–36.

[43] Hanik, N., et al., "Optimized Design of Transparent Optical Domains," *ECOC'2000 Proc.*, Vol. 3, Munich, 2000, pp. 195–197.

[44] Palais, J. G., T. Y. Lin, and S. Tariq, "Power Limitations in Fiber-Optic Frequency Division Multiplexed Systems," *Fiber and Integrated Optics*, Vol. 10, London: Taylor and Francis Ed., 1991, pp. 75–94.

[45] Aoki, Y., K. Tajima, and I. Mito, "Input Power Limits of Single-Mode Optical Fibers Due to Stimulated Brillouin Scattering in Optical Communication Systems," *J. of Lightwave Technol.*, Vol. 6, No. 10, 1998, pp. 710–719.

[46] Tkack, R. W., A. R. Chraplyvy, and R. M. Derosier, "Spontaneous Brillouin Scattering for Single-Mode Optical-Fibre Characterization," *Electron. Lett.*, Vol. 22, 1986, p. 1011–1013.

[47] Yeniay, A., J. M Delavaux, and J. Toulouse, "Polarization Multiplexing Technique for SBS Suppression," *ECOC'2000 Proc.*, Vol. 3, Munich, 2000, pp. 91–92.

[48] Franck, T., T. Norskov, and A. Stentz, "Experimental Verification of SBS Suppression by Duobinary Modulation," *ECOC'97 IEE Proc.*, Vol. 1, Edinburg, 1997, pp. 71–74.

[49] Smith, R. G., "Optical Power Handling Capacity of Low Loss Optical Fibers as Determined by Stimulated Raman and Brillouin Scattering," *Applied Optics*, Vol. 11, No. 11, Nov. 1972, pp. 2489–2494.

[50] Chraplyvy, A. R., and P. S. Henry, "Performance Degradation Due to Stimulated Raman Scattering in Wavelength Division Multiplexed Optical Fiber Systems," *Electron. Lett.*, Vol. 19, 1983, pp. 641–643.

[51] Chraplyvy, A. R., "Optical Power Limits in Multi-Channel Wavelength Division Multiplexed Systems Due to Stimulated Raman Scattering," *Electron. Lett.*, Vol. 20, 1984, pp. 58–59.

[52] Cotter, D., and A. M. Hill, "Stimulated Raman Crosstalk in Optical Transmission: Effects of Group Velocity Dispersion," *Electron. Lett.*, Vol. 20, 1984, pp. 185–187.

[53] Jiang, W., and P. Ye, "Crosstalk in Fiber Raman Amplification for WDM Systems," *IEEE J. Lightwave Technol.*, Vol. 7, 1989, pp. 1407–1411.

[54] Chraplyvy, A. R., "Limitation on Lightwave Communications Imposed by Optical Fiber Nonlinearities," *IEEE J. Lightwave Technol.*, Vol. 8, 1990, pp. 1548–1557.

[55] Krummrich, R. E., et al., "Influence of Stimulated Raman Scattering on the Channel Power Balance in Bi-Directional WDM Transmission," *OFC 99 Proc.*, WJ6, San Diego, CA, Feb. 21–26, 1999, pp. 171–176.

[56] Krummrich, R. E., P. M. Neuhauser, and G. Fisher, "Experimental Comparison of Raman Thresholds of Different Transmission Fibre Type," *ECOC'2000 Proc.*, Vol. 3, Munich, 2000, pp. 133–134.

[57] Niay, P., et al., "Mise en Évidence de la Diaphonie Raman dans un Système de Transmission Analogique à Multiplexage Chromatique sur Fibre Monomode," *Opto'98 Proc.*, No. 47, Paris: ESI Ed., Nov. 1988, pp. 42–47.

[58] Bertaina, A., and D. Chiaroni, "Optical WDM Transmissions: State of the Art, Limitations, Perspectives," *OPNET'99* Actes, ENST, Paris, Jan. 26–28, 1999, pp. 113–133.

[59] Chi, S., and S. C. Wang, "Maximum Bitrate Length Product in the High Density WDM Optical Fibre Communication System," *Electron. Lett.*, Vol. 26, No. 18, August 30, 1990, pp. 1509–1512.

[60] Hegarty, J., N. A. Olsson, and M. McGlashan-Powell, "Measurement of the Raman Crosstalk at 1,5 μm in a Wavelength Division Multiplexed Transmission System," *Electron. Lett.*, Vol. 21, 1985, pp. 395–397.

[61] Cotter D., "Stimulated Brillouin Scattering in Monomode Fibre," *J. Opt. Com.*, Vol. 4, 1983, pp. 10–19.

[62] Waarts, R. G., and R. P. Braun, "Crosstalk Due to Stimulated Brillouin Scattering in Monomode Fibre," *Electron. Lett.*, Vol. 21, 1985, pp. 1114–1115.

[63] Comuzzi, R., C. De Angelis, and G. Gianello, "Improved Analysis of the Effects of Stimulated Raman Scattering in a Multi-Channel WDM Communication System," *Eur. Trans. Telecom. Relat. Technol. (Italy)*, Vol. 3, No. 3, May/June 1992, pp. 295–298.

[64] McIntosh, C. M., et al., "Threshold Enhancement in a Massive WDM System Via Optical Filtering Techniques," *ECOC'2000 Proc.*, Vol. 3, Munich, 2000, pp. 41–42.

[65] Yano, Y., et al., "Experimental Study on SRS Loss and Its Compensation in Three-Band WDM Transmission," *ECOC'2000 Proc.*, Vol. 3, Munich, 2000, pp. 39–40.

[66] Keang-PoHo, "Statistical Properties of Stimulated Raman Crosstalk in WDM Systems," *IEEE J. Lightwave Technol.*, Vol. 18, No. 7, 2000, pp. 915–921.

[67] Le Roux, P., et al., "Error-Free 2.5 Gbit/s Unrepeatered Transmission over 570 Km," *ECOC'2000 Proc.*, Vol. 4, Munich, 2000, pp. 45–46.

[68] Krummrich, et al., "40 Gb/s ETDM for Long Haul WDM Transmission," *ECOC'2000 Proc.*, Vol. 4, Munich, Sept. 3–7, 2000, pp. 13–14.

[69] Bigo, S., et al., "1.28 Tbit/s WDM Transmission of 32 Channels at 40 Gbit/s over 3 × 100 Km Distance," *ECOC'2000 Proc.*, Vol. 4, Munich, 2000, pp. 17–18.

[70] Miyamoto, Y., et al., "1.2 Tbit/s (30 × 42.7 Gbit/s ETDM Optical Channel) WDM Transmission over 376 Km with 125 Km Spacing Using Forward Error Correction and Carrier-Suppressed RZ Format," *OFC'00 Proc.*, PD26, Baltimore, MD, 2000.

[71] Nielsen, et al., "3.28 Tb/s (82 × 40 Gb/s) Transmission over 3 × 100 Km Non-Zero-Dispersion Fibre Using Dual C and L Band Hybrid Raman/Erbium Doped in Line Amplifiers," *OFC'00 Proc.*, Post deadline papers 23, Baltimore, MD, 2000.

[72] Elbert, J. P., et al., "3.2 Tbit/s (80 × 40 Gbit/s) Bi-Directional DWDM/ETDM Transmission," *ECOC'1999 Proc.*, Post deadline papers 2–5, Nice, France, 1999.

[73] Frignac, et al., "Numerical Optimization of Residual Dispersion in Dispersion-Managed Systems at 40 Gb/s," *OFC'00 Proc.*, TuD3, Baltimore, MD, 2000.

[74] Brandon, E., et al., "1.28 Tbit/s (32 × 40 Gbit/s) Unrepeated Transmission over 250 Km," *ECOC'2000 Proc.*, Vol. 4, Munich, Sept. 3–7, 2000, pp. 21–22.

[75] Morita, I., et al., "40 Gbit/s × 16 WDM Transmission over 2000 Km Using Dispersion Managed Low Linear Fiber Span," *ECOC'2000 Proc.*, Vol. 4, Munich, Sept. 3–7, 2000, pp. 25–26.

[76] Pratt, A. R., et al., "8 × 40 Gbit/s over 640 Km of Large Effective Area Nonzero-Dispersion Shifted Fiber," *ECOC'2000 Proc.*, Vol. 4, Munich, Sept. 3–7, 2000, pp. 29–30.

[77] Ito, T., et al., "6.4 Tb/s (160 × 40 Gbit/s) WDM Transmission Experiment with 0.8 bit/s/Hz Spectral Efficiency," *ECOC'2000 Proc.*, Post-deadline papers 1.1, Munich, Sept. 3–7, 2000.

[78] Bigo, S., et al. "5.12 Tb/s (128 × 40 Gbit/s WDM) Transmission over 3 × 100 Km of Teralight TM Fiber," *ECOC'2000 Proc.*, Post-deadline papers 1.2, Munich, Sept. 3–7, 2000.

[79] Färbert, A., et al., "7 Tb/s (178 × 40 Gbit/s WDM) Bi-Ddirectional Interleaved Transmission with 50 GHz Spacing," *ECOC'2000 Proc.*, Post-deadline papers 1.3, Munich, Sept. 3–7, 2000.

[80] Zhu, Y., et al., "1.28 Tb/s (32 × 40 Gbit/s) Transmission over 1000 Km with Only 6 Spans," *ECOC'2000 Proc.*, Post-deadline papers 1.4, Munich, Sept. 3–7, 2000.

[81] Tsuritani, T., et al., "35 GHz-Spaced 20 Gbit/s × 100 WDM RZ Transmission over 2700 Km Using SMF-Based Dispersion Flattened Fiber Span," *ECOC'2000 Proc.*, Post-deadline papers 1.5, Munich, Sept. 3–7, 2000.

[82] Cai, J-X., et al., "1.12 Tb/s Transmission over Trans-Atlantic Distance (6200 Km) Using Fifty-Six 20 Gbit/s Channels," *ECOC'2000 Proc.*, Post-deadline papers 1.6, Munich, Sept. 3–7, 2000.

[83] Kobayashi, Y.. et al., "A Comparison Among Pure-RZ , CS-RZ and SSB -RZ Format, in 1 Tb/s (50 × 20 Gbit/s, 0,4 nm Spacing) WDM Transmission over 4000 Km," *ECOC'2000 Proc.*, Post-deadline papers 1.7, Munich, Sept. 3–7, 2000.

[84] Tanaka, T., et al., "2.1 Tb/s WDM Transmission over 7221 Km with 80-Km Repeater Spacing," *ECOC'2000 Proc.*, Post-deadline papers 1.8, Munich, Sept. 3–7, 2000.

[85] Jenkins, R. B., "Cross-Phase Modulation and Multiwavelength Adiabatic Solitons," *Optics Lett.*, Vol. 25, Issue 9, May 1, 2000, pp. 604–606.

[86] Hill, K. O., et al., "CW Three-Wave Mixing in Single-Mode Fibers," *J. Appl. Phys.*, Vol. 49, Oct. 1978, pp. 5098–5106.

[87] Atkinson, M. D., N. Santoro, and J. Urrutia, "Integer Sets with Distinct Sums and Differences and Carrier Frequency Assignments for Nonlinear Repeater," *IEEE Trans. Comm.*, Vol. COM-34, June 1986, pp. 614–617.

[88] Forghieri, F., et al., "Reduction of Four Wave Mixing Crosstalk in WDM Systems Using Unequally Spaced Channels," *IEEE Photon Technol. Lett.*, Vol. 6, No. 6, pp. 754–756, June 1994.

[89] Tkach, R. W., et al., "Four-Photon Mixing and High-Speed WDM Systems," *IEEE J. Lightwave Technol.*, Vol. 13, No. 5, May 1995, pp. 841–849.

[90] Forghieri, F., R. W. Tkach, and A. R. Chraplyvy, "WDM Systems with Unequally Spaced Channels," *IEEE J. Lightwave Technol.*, Vol. 13, No. 5, May 1995, pp. 889–897.

[91] Miao, X., "Unequally Spaced Channels for Upgrading WDM Systems from 3-Channels to 22-Channel Preserving No FWM Crosstalk," *Applications of Photonic Technology: Communications, Sensing, Materials and Signal Processing*, Vol. 2, 1997, pp. 391–397.

[92] Hjelme, D. R., J. Eide, and B. J. Slagsvold, "Four-Wave-Mixing Efficiency in Installed Dispersion Shifted Fibers: Effects of Zero-Dispersion Wavelength Variations," *OFC'98 Proc.*, Vol. 2, San Jose, CA, 1998, pp. 209–210.

[93] Shake, et al., "Influence of Inter-Bit Four-Wave Mixing in Optical TDM Transmission," *Electron. Lett.*, Vol. 34, No. 6, Aug. 6, 1998, pp. 1600–1601.

[94] Wemer, D., A. Schinabeck, and B. Schmauss, "Limits in Optical Long Haul Transport Networks Caused by OSNR, SPM, and Channel Spacing," *ACTS Workshop on WDM Networks and Related Technologies Proc.*, Brussels, May 6, 1998, pp. 12–22.

8

Application of DWDM to Telecommunication Networks

8.1 Some of the Earlier Applications

8.1.1 Introduction

The first WDM applications to customer-premise networks (CPN) appeared in the early 1980s. The first networks used multimode fibers and very often two wavelengths, 0.8 and 1.3 μm, or sometimes three or four wavelengths in WDM configurations. It soon became obvious that the use of SM fibers was necessary to achieve the bandwidth required on high bit-rate trunks. In the 1980s, the great advantages of SM fibers for local networks, when switching, routing, and processing of signals had to be done in the optical domain, were increasingly considered. Several groups were pursuing plans to increase the number of multiplexed wavelengths and, should the occasion arise, to use wavelength tunable sources and receivers. A better understanding emerged of the fact that optical networks could remain totally passive and, for instance, that switching and routing could use the wavelength at the terminal level. In the following part of Section 8.1 we will present only a few typical networking examples published over the last two decades. This does not mean that other examples not mentioned would be neither less interesting nor less important; they are simply too numerous to mention them all.

8.1.2 First Broadband Multiwavelength Passive Optical Networks

Typically, different stations emitted specific wavelengths into the optical network, through a central star coupler or a bus, for instance. All optical signals emitted in the networks reached each station. A wavelength division multiplexer at each station was able to separate in parallel the different wavelength signals from the different network emitters. In general, the routing was electronic as in "classical networks," using time switching with a selection done through the address coded on the message header. These networks demonstrated many advantages, in particular, few or no blocking possibilities and either multipoint-to-multipoint link, point-to-point link, or point-to-multipoint link.

Between 1985 and 1990, many demonstrations confirmed that bandwidth × distance products larger than 1.5 $Tbps^{-1}$ km were obtained easily without repeaters, with the components already available at that time, in broadcast and select networks using star structures.

- In 1985, British Telecom Research Laboratories (BTRL) introduced a new wavelength switched local network [1]. The network terminals were interconnected through a central star coupler. A specific wavelength emitter, as well as a tunable wavelength filter receiver, was allocated to each terminal. A network control station operated at its own allocated wavelength, polling each terminal in turn. In the free state, the stations were tuned to the control wavelength. Polling and selecting any terminal on a given optical wavelength was under the control of this station. Other versions were proposed in which each terminal had a tunable rather than a fixed wavelength optical source.
- In 1986, Bell Communication Research introduced the Lambdanet network [2], interconnecting 16 stations through a "transparent," SM, optical-fiber star network. A 16 × 16 star coupler was used; each station was emitting its own wavelength and receiving all the other wavelengths on a 16-channel WDM.
- Published a few years later, the WTDM program managed by the BBC (U.K.) was another application of a similar passive star network to the development of a broadband customer-premises network (BCPN), for the internal routing network of a television studio center [3, 4].
- In 1991, another TV studio network had also been built in Japan [5].

- Reported in 1988, the Bell Communication Research's Passive Photonic Loop (PPL) network is another example of a star structure. This network used a double passive star of SM fibers between the central office (CO) and the remote terminals (RT) [6]. In such a network, the switched services to the different subscribers were multiplexed on a fiber installed between the CO and the RTs. The signals were switched and routed to their intended destinations in a terminal. In the direction RT to CO, the reverse occurred. These functions could be fulfilled using time multiplexing exclusively, but with sophisticated electronics in the RT. WDM eliminated this drawback and made it possible to build such a network with passive optical components.

8.1.3 One of the First Access Networks Using a Multiwavelength Passive Tree

Studied before 1990, the British Telecom's broadband passive optical network (BPON) proposed an SM passive tree structure between the exchange and the customer terminals. Prior to any other consideration, the economic justification took into account the basis of known revenue earnings from telephone services, but the network was prepared for a later requirement for broadband services that will use other wavelengths. These networks fully illustrated early on one of the most interesting characteristics of wavelength division multiplexing: It allows network evolution without questions about the infrastructure [7, 8]. A wavelength filter was included first in the customer equipment for the selection of the telephone service. The other services, and in particular the broadband services, could be added on the other wavelengths later on. In these networks, the time-multiplexed signals were broadcast from the exchange to all customers through several passive optical splitters. The whole system involved time management and switching that was controlled by the customer destination encoded on the data prior to transmission. As in all telecommunications systems, the quality of the time management was one of the key features of the system.

8.1.4 Some of the Earlier DWDM Virtual Topologies

In a logical-topology multiwavelength architecture, electronic logical network overlays are used over the optical network. The topology is called virtual when it is concerned only with logical connections between stations. One example of an optical multiplexing application is to create virtual topologies on request so that the network configuration can be modified independently of its physical topology by modification of the emitted or received optical frequencies. The Shuffle Net, a multiwavelength bus [9, 10] and the WON [11], a wavelength-division optical network that is a generalization of ShuffleNet, were early interesting examples of this ability to create a virtual topology adjustable through wavelength division multiplexing. ShuffleNet was a unidirectional optical bus connecting a set of network interface units (NIU), each with two fixed emission wavelengths and two fixed receiving wavelengths. This architecture could be applied to packet switching. A message from any NIU to any other NIU could be addressed. In each unit, the transmitter and receiver wavelengths were necessarily different; for example, 16 wavelengths were used for eight NIUs. They were interleaved as in a shuffled pack of cards. It was possible to go from one NIU to another NIU either directly or by multihops at different wavelength of the data packet with its own address code. The path could be modified: In case of congestion, an alternative path of hops could be used. The design was proposed for MAN implementation. In the WON network the product propagation delay × throughput was optimized. The concept was later generalized to the control of other types of optical networks.

8.2 Today's DWDM Networks

8.2.1 Introduction

At the present time, telecommunication networks are relatively heterogeneous. This is the consequence of an evolution, during the course of which the transmission hardware as well as the operation methods were progressively modified to meet ever-changing service requirements.

New services have to be introduced by traditional and new operators. We see a tremendous growth of the Internet. New services such as video conferencing, entertainment, interactive video training, teleworking, bank information service, data transfer between computers, teledistribution, and generally all ISDN, and broadband networks, need or will need bit rates from 1 to 100 Mbps and above. The new networks require a real flexibility and a kind of distributed intelligence.

8.2.2 Networks' Topologies and Transfer Modes

The different network types may be sorted according to their size, their topology, or according to their transfer mode. However, the limits of the different categories are not always well defined and a given network may superpose different topologies and formats.

According to the network size, the following terms are often used:

- Local area network (LAN);
- Metropolitan area network (MAN);
- Access network;
- Wide area network (WAN);
- Transport network.

According to their topologies, the networks are called:

- Static broadcast and select networks;
- Wavelength routed networks;
- Linear lightwave networks;
- Logically routed networks.

The main standard transfer modes are:

- IP: IP control cards read the destination in the packet header and transmit the packet of data in the next available time-slot in the direction given as the most appropriate. The packets have variable lengths.
- Synchronous digital hierarchy (SDH): Signals arranged with the data in 810-byte frames with an header of 18 bytes of networking information used in circuit-switched networks optimized in the past for voice traffic. The line rate is in hierarchies: STM-1 (155.84 Mbps), STM-2 (622.08 Mbps), STM-16 (2488.32 Mbps, STM-64 (9953.28 Mbps).
- SONET: Like SDH, SONET was designed primarily for digital telephone optical network transmission with a centralized management system. Like SDH it uses standard time division multiplexing with optical carriers (OC) at hierarchical bit rates from OC-1 (51.48 Mbps), OC-2 (155.52 Mbps) to OC-192 (9953.84 Mbps). Standards for maintenance and protection are included. The quality of service is better than that of IP, but the use of resources is less effective.

- Asynchronous transfer mode (ATM): ATM was developed for broadband ISDN (B-ISDN). It uses packets of uniform length of 53 bytes.
- Ethernet: This format was developed for local-area networks. It includes "basic," "fast," and "Gigabit"' Ethernet at 10 Mbps, 100 Mbps and 1 Gbps, respectively.
- Simple data link (SDL).
- Multiprotocol label switching (MPLS): MPLS can adapt data of any type into one packet format. It does not need a centralized management. It has a set of protocols simpler than in ATM. In fact, MPLS was primarily developed for IP networks. It implements electronically label-switched paths (virtual connections) in the network through label-switching routers. Multiprotocol lambda switching (MPλS) is a current extension of MPLS to manage optical-network connections.
- Frame relay: Packet-based data-oriented protocol typically used for interconnection of local-area networks.

8.2.3 WDM for Optical Switching and Routing

Traditionally, network architectures are schematically stacked in different layers (Figure 8.1). We have, at the upper levels: services and applications layers; at the intermediate levels: electrical switch/multiplex layers such as an ATM layer on a SDH/SONET layer; and at the first level the transport network with its DWDM optical transmission layer. In a network node, the function of the switch is to connect an input channel to a specified output channel in such a way that the output signals are as close as possible to the input signals.

The devices allowing for this function traditionally use space and time switches in the electrical layers. Today, most of the switching and routing operations are done in the electronics layers. In the ATM electronic layer, ATM cross-connect/switches operate on the ATM cells. In the SDH/SONET electronic layer the digital cross-connects (DXC) operate on semipermanent synchronous virtual containers (VC). Time switching is linked to time division multiplexed digital transmissions. In each of the channels of such a multiplex, a time window is allocated. The switch modifies the time position of the entrance signal and ensures that the corresponding information train between multiplexes is synchronous.

In principle, in the photonic WDM networks, space, time, and wavelength switching can be added. Thus, a new dimension, allowing for new capabilities and flexibility, becomes available. Interestingly, the aggregate capacity can be increased and the blocking probability (i. e., the probability that all possible routes are already in use at a given time), can be reduced. Optical switching, in a first step through OADMs in ring networks, and progressively with larger and larger optical routers and cross-connects (OXC) in interconnected ring and meshed networks, allows for optical protection, path restoration, and/or network reconfiguration in the photonic transport layer. In many cases, the current IP-backbone is provided by SONET/SDH systems on point-to-point WDM links. OADMs and OXCs with fixed wavelengths are already available and just beginning to be deployed in the network nodes. They help avoid unnecessary O-E-O conversions, bypassing as much as possible in the optical layer all traffic that does not directly concern the node (Figure 8.1).

8.2.4 Evolution Towards IP Over DWDM

In the beginning of the twenty-first century, the well-standardized SDH/SONET protocol still dominated in medium- and long-haul transmissions that were optimized for voice traffic. As new networks are aggressively deployed, the other standard protocols are also employed according to the specific needs. IP becomes more and more the dominant technology.

WDM and DWDM, that is of course compatible with TDM, is used in networks of all size, but now predominantly in WANs. WDM/DWDM, compatible with any transfer mode, is not fully standardized. With a granularity of 10 Gbps per channel, one can envisage future standard hierarchies corresponding to a capacity of 320 Gbps with 32 wavelengths, to 1.6 Tbps with 160 wavelengths. Already higher capacities are obtained at the laboratory stage. The management of IP over DWDM seems one of the most interesting options for the near future [12]. In principle, it allows dealing with hundreds of OC-192 channels in the optical layer. It can be implemented with an intermediate simple Ethernet layer or with a superposition of different layers using SONET, ATM, SDL, MPLS, and/or other options.

8.2.5 Optical Packet Switching

In the future, active DWDM may evolve towards a true fast optical switching, such as optical packet switching [13–16]. However, at the present time,

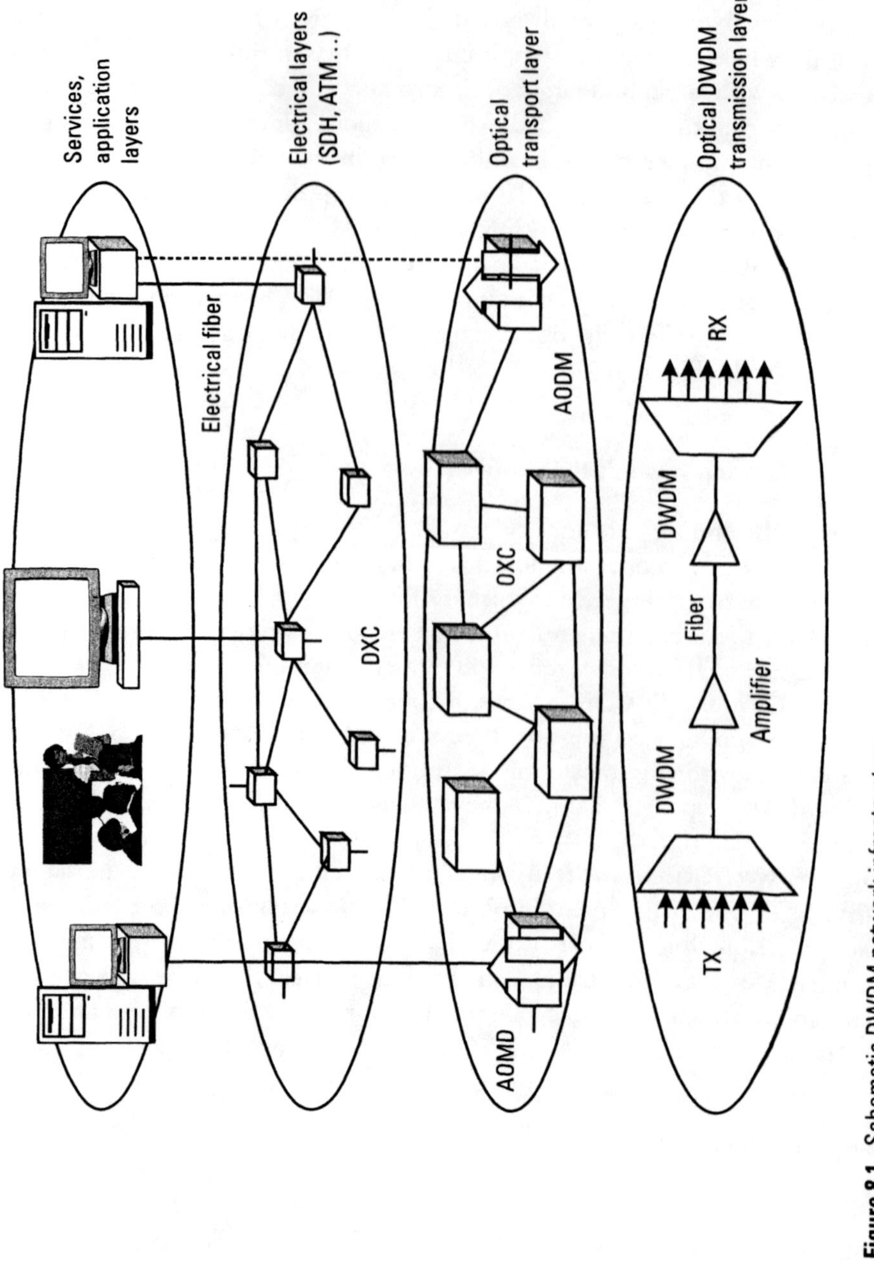

Figure 8.1 Schematic DWDM network infrastructure.

the lack of adequate optical memories and optical programmable delay lines is still a problem delaying the introduction of full optical switching.

An optical packet network consists of optical packet switches interconnected with WDM links. The user data is transmitted in packet payload, which are switched entirely in the optical domain. The payload doesn't experience optical-to-electrical or electrical-to-optical conversion between source and destination. The move from circuit-based switching towards packet-based switching has been a general trend in datacom and telecom networks for many years. Indeed, by allowing the sharing of the available bandwidth with a very-fine granularity, using packets results in much better bandwidth efficiency. This becomes particularly true in modern telecom networks, where Internet traffic becomes largely predominant. Moreover, a significant increase in statistical multiplexing gain may be achieved by considering the entire fiber bandwidth, or at least parts of this bandwidth, as a shared resource, instead of considering a WDM link a collection of independent transmission channels. A further advantage of optical packet switching is bit-rate transparency, giving the possibility to support, with minor impact on the switching nodes, successive increases of the transmission link bit rate. This approach could lead to dissociate efficiently transmission/switching functions (carried in optical domain to access the overall fiber bandwidth) and routing/forwarding functions (based on header processing in the electronic domain). Further information on optical packet switching may be found in [17–20].

Packet switching not only includes purely optical packet switches (according to our previous definition) but also schemes, such as hybrid packet switches (with electronic buffers) or burst switches (possibly without buffers). Optical burst switching (OBS) [21, 22] is a technique for transmitting bursts of traffic through an optical transport network by setting up a connection and reserving end-to-end resources for the duration of the burst. Buffering at the WDM layer is not required. For instance, OBS facilitates IP-over-WDM integration under the MPLS protocol [23].

In the optical packet node, besides the electronic control subsystem, the main blocks are the packet-synchronization interfaces, a fast space-switch, and a fiber delay line-based buffer (not necessary in hybrid packet switching or in optical burst switching) allowing a solution to contention between packets targeted towards the same output port. High-capacity wavelength routers, fast tunable lasers, and fast optical space-switching matrices are key enabling devices for the realization of optical packet switches. Large routers (48 × 48 and more, see Chapter 6) are already commercially available. Lasers tunable over 60 nm with less than 5 ns switching times (See

Chapter 4 and [24]), and switches based on such technology as SOAs, or electroholography with 10-ns response times (physical limit down 1 ns, see Chapter 6 and [25]) are becoming available. This means that optical packet-switching networks, in particular using MPλS [26], can be designed for the next-generation optical Internet.

8.3 Long-Distance Transmission

In this section we will limit our comments to systems used in ultra-long-distance submarine and moderate-length terrestrial DWDM transmissions.

Some of the characteristics of different systems, published at the time of this writing, are compared in Table 8.1 [27–36]. Developments in this area are moving so fast that the practical solutions the installed links will take are still not clear. Also, we could not be exhaustive, other important and good results from other laboratories and other manufacturers could not be reported in this chapter. More fundamental aspects of soliton transmissions are presented in Chapter 2.

The ultimate capacity will depend on the spectral efficiency defined as the aggregate capacity divided by the optical bandwidth of the system (in bps/Hz) [27]. To our knowledge, in the best ultra-long-distance experimental systems the spectral efficiency is typically 0.2 to 0.4 bps/Hz. The spectral efficiencies can be about 1 bps/Hz for shorter distances (for instance 0.8 bps/Hz, 160 km in [35]).

Table 8.1 illustrates how the addition of Raman amplification to EDFA amplification allowing for an increase of the span length (up to 160 or 200 km at 40 Gbps in [34]). A very dense channel spacing can be used. A channel spacing as low as 25 GHz was proposed in advanced systems using soliton transmissions. Traditionally, dispersion-managed solitons have allowed the longest transmissions at 10 to 80 Gbps without regeneration [37–39]. 16,000 km at 10 Gbps and 1,000 km at 40 Gbps with a standard fiber have been obtained. DWDM transmissions using nonlinearity-controlled RZ formats (in particular dispersion-managed solitons) allow for multiterabit capacities beyond 1,000 km. Soliton or quasi-soliton (NRZ) solutions plus EDFA/Raman amplification have been implemented in all-optical ultra-long-haul transmissions by several carriers. However, other solutions, with other formats, with or without Raman amplification, could be preferred sometimes. Use of a spectrum-broadening modulation format like chirped RZ is effective [40]. For instance the full-spectrum WDM [41] technology allows for the launching of signals at higher average power per

Table 8.1
Typical Characteristics in Ultra-Long-Distance DWDM Transmission

Reference ***Number of Channels***	**Aggregate Capacity Tb/s** ***Bit rate Gb/s***	**Distance Total Km** ***Span Km***	**Fiber Type**	**Wavelength Band** ***Amplifiers***	**Channel Spacing GHz**	**Modulation Format** ***Correction***
[30]	2.1	7,221	Positive and negative-dispersion fibers	C + L, EDFA + Distributed Raman	37.5	RZ
211	10	80				FEC
[27]	1.8	7,000	not given	C	About 33	Chirped RZ
180	10	not given		not given	not given	FEC
[32]	1.12	6,200	LEA	C	85 to 170	Chirped RZ
56	20	49	RDF			FEC
[28]	1	4,000	NZDSF+ LEA	C + L	44	Amplitude NRZ
100	10	48	+ PSCF+ RDF	EDFA		+ Intensity RZ shaping, FEC
[31]	1	4,000	SMF +		50	RZ, CS-RZ,
50	20	41.5	RDF	EDFA		SSB-RZ (CS-RZ preferred)
[29]	0.8	3,200	Low dispersion slope TrueWave + DCF	C	50	NRZ
80	10	100		EDFA + Raman		FEC
[33]	2	2,700	SMF based PDF + SCF	C	35	RZ
100	20	39.5				not given
[34]	1.28	1,000	NDSF	2 stage EDFA + Raman	100	CS-RZ
32	40	160	DCM			not given
[36]	5.1	300	TeraLight	EDFA	50 and 75 alternate	NRZ
128	40	100	DCF	Raman		VSB filtering
[35]	6.4	186	SMF,RDF,DCF	C + L	50	NRZ
160	40	46.6		EDFA		*Polarization interleaving*
[37]	1 (4)	4,000	Different fibers with dispersion management	Extended C (C + L) EDFA Raman	50	Soliton
100	10	100				FEC
[39]	3.2 (2.8)	800 (3,200)	Different fibers with dispersion management	C + L	25	Soliton
320	10	not given				Not given
[41–43]	0.4 (1.6)	4,000 (6,800)	Various types	C , (C + L)	100	RZ, FSWDM
40 (160)	10	80–120		Special EDFA	(50)	FEC

NZDSF: Non-zero dispersion shifted fiber
LEA: Large effective area fiber ($70\mu m^2$)
SCF: Slope compensating fibers
PSCF: Positive slope compensating fibers
RDF Reverse dispersion fiber
DCF: Dispersion-compensating fiber
DCM: Dispersion compensation module
EDFA: Erbium doped fiber amplifier
NRZ: Non-return-to zero
FEC: Forward error correction
FSWDM: Full-Spectrum WDM

channel without an increase of FWM and cross-modulation, as a wider spectrum in each channel (typically 30 to 40 GHz width for 100-GHz-channel spacing) reduces the spectral density. The accumulated effects of self-phase modulation are compensated with appropriate control of frequency and time-domain modulations. Such systems are claimed to be "capable of propagating signals over distances of 6,800 km in SMF without Raman amplification and without precise span-by-span dispersion tuning" [42, 43]. Sometimes, conversely, it can be interesting to reduce the channel spacing with a spectral-width-reducing modulation format in order to increase the total transmission capacity. For this the carrier-suppressed return-to-zero (CS-RZ) modulation format is unique to have a relatively high nonlinearity tolerance with a narrow spectrum [44]. In CS-RZ modulation the optical pulses have a phase 0 or π alternatively, which reduces the pulses interactions [45].

8.4 Other DWDM Applications

At the present time, DWDM is also efficient in analog video transport networks (such as hybrid fiber/coax CATV/data networks [46]). DWDM is compatible with broadband distribution using subcarrier multiplexing (SCM), that is, an electronic frequency modulation technique. DWDM is also used in various specific networks such as sensor networks, remote radar networks, and telespectroscopic process control networks. Many examples can be added to each of the above categories.

8.5 Conclusion

The transmission and, above all, the switching of a larger and larger number of signals that are increasingly diversified, and with aggregate bandwidths now larger than 1 Tbps for several applications, lead to a bottleneck on the lines and electronic circuits which only optical transmission and processing can relieve. DWDM allows for a linear increase in the number of channels; this number having its own limits, namely the total power that can be transmitted in the fiber without redhibitory nonlinear effects or damages. It contributes to more flexibility in system design and, moreover, it allows new switching and routing facilities in the optical domain. Besides, the optical multiplexer allows for the evolution of existing systems and can contribute to incredible network cost saving.

References

[1] Payne, D. B., and J. R. Stern, "Single Mode Optical Local Networks," Conf. Proc. *GlobeCom'85*, Houston, TX, 1985, Paper 39.5.

[2] Goodman, M. S., H. Kobrinski, and K. W. Loh, "Application of Wavelength Division Multiplexing to Communication Networks Architectures," *ICC'86, Conference Record*, Toronto, 1986, p. 931.

[3] Marsden, R. P., et al., "A Multi-Gigabit/s Optical Business Communication System Using Wavelength and Time Division Multiplexing Techniques," *ECOC'90 Proc.*, Vol. 2, 1990, pp. 779–786.

[4] Newell, J. C., "The Use of an Optical Amplifier for Extending the Transmission Distance in a WDM Network," *SG1-Optical Networks Workshop Proc.*, Brussels (RACE), Jan. 13–14, 1993.

[5] Shimosaka, N., et al., "A Photonic Wavelength-Division and Time-Division Hybrid Multiplexed Network Using Tunable Wavelength Filters for a Broadcasting Studio Application," *ECOC'91 Proc.*, 1991, pp. 545–548.

[6] Wagner, S., et al., "Experimental Demonstration of a Passive Optical Subscriber Loop Architecture," *Elect. Lett.*, Vol. 24, March 1988, pp. 325–326.

[7] Stern, J. R., et al., "Passive Optical Networks for Telephony Application and Beyond," *Electron. Lett.*, Vol. 23, No. 24, Nov. 19, 1988, pp. 203–206.

[8] Hoppitt, C. E., and D. E. Clarke, "The Provision of Telephony over Passive Optical Network," *British Telecom. Technol. J.*, Vol. 7, No. 2, April 1989, pp. 100–114.

[9] Acampora, A. S., "A Multichannel Multihop Local Lightwave Network," *Globecom'87 Proc.*, 1987, pp. 1459–1467.

[10] Hluchyjand, M. G., and M. J. Karol, "Shuffle Net: An Application of Generalized Perfect Shuffles to Multihop Lightwave Networks," *IEEE J. Lightwave Technol.*, Vol. 9, No. 10, Oct. 1991, pp. 1386–1397.

[11] Bannister, J. A., and M. Gerla, "Design of the Wavelength-Division Optical Network," *ICC 90, IEE*, Vol.3, April 1990, pp. 962–967.

[12] Jaeger, M., H. -M. Foisel, and F. -J. Westphal, "IP over WDM Evolution," *ECOC'2000*, Vol.2, Munich, September 5, 2000, pp. 83–86.

[13] Takada, A., et al., "Ultrafast Packet Ring Network Employing Optical Label Switching," *ECOC'2000*, Vol.2, Munich, September 5, 2000, pp. 19–20.

[14] Le Sauze N., et al., "Modular Optical Packet Switching Node for Future Multi-QoS Metro Networks," *ECOC'2000*, Vol.2, Munich, September 5, 2000, pp. 20–21.

[15] Düser, M., et al., "Design Trade-Offs in Optical Burst Switched Networks with Dynamic Wavelength Allocation," *ECOC'2000*, Vol.2, Munich, September 5, 2000, pp. 23–24.

[16] Laude, J. P., B. Fracasso, and P. Gravey, "Applications of High Capacity Wavelength Routers" *OFC'2001, Application note, HighWave,* March 2001.

[17] Blumenthal, D. J., P. R. Prucnal, and J. R. Sauer, "Photonic Packet Switches: Architectures and Experimental Implementations," *Proc. of the IEEE,* Vol. 82, No. 199411, pp. 1650–1667.

[18] Tucker, R. S., and W. De Zhong, "Photonic Packet Switching: An Overview," *IEICE Trans. Electron.*, Vol. E82-C, No. 2, 1999, pp. 202–212.

[19] Gambini, P., et al., "Transparent Optical Packet Switching: Network Architecture and Demonstrators in the KEOPS Project," *IEEE J. Selected Areas Comm.*, Vol. 16, No. 7, 1998, pp. 1245–1259.

[20] Dixit, S., and P. Lin, "Advances in Packet Switching/Routing in Optical Networks," *IEEE Communications Magazine,* Feb. 2001, p. 79.

[21] Qiao, C., and M. Yoo, "Optical Burst Switching (OBS)—A New Paradigm for an Optical Internet," *J. High Speed Networks,* Vol. 8, No. 1, 1999, pp. 69–84.

[22] Yoo, M., C. Qiao, and S. Dixit, "Optical Burst Switching for Service Differentiation in the Next-Generation Optical Internet," *IEEE Communications Magazine,* Feb. 2001, pp. 98–104.

[23] Awduche, D., "MPLS and Traffic Engineering in IP Networks," *IEEE Communications Magazine,* Dec. 1999.

[24] Lavrova, O. A., G. Rossi, and D. J. Blumenthal, "Rapid Tunable Transmitter with Large Numbers of ITU Channels Accessible in Less Than 5 ns," *ECOC 2000 Proc.*, Vol. 2, 2000, pp. 169–17.

[25] Cahall, T., "Holographic Switches Can Open up the Bottlenecks," *FibreSystems Europe,* May 2001, pp. 50–52.

[26] Awduche, et al., "Multi-Protocol Lambda Switching: Combining MPLS Traffic Engineering Control with Optical Crossconnects," *Internet Engineering Task Force* (IETF) *Internet draft,* Nov. 1999.

[27] Bergano, N., "Ultra Long Distance Submarine DWDM Systems," *ECOC'2000 Proc.*, Vol. 1, Munich, Sept. 4, 2000, pp. 67–68.

[28] Vareille, G., et al., "1 Tbit/s WDM C + L Band Transmission over 4,000 km of NonZero Dispersion Shifted Fiber," *ECOC'2000 Proc.*, Vol. 1, Munich, Sept. 4, 2000, pp. 69–70.

[29] Zhu, B., et al., "800 Gb/s NRZ Transmission over 3,200 km of Truewave Fiber with 100-km Amplified Spans Employing Distributed Raman Amplification," *ECOC'2000 Proc.*, Vol. 1, Munich, Sept. 4, 2000, pp. 73–74.

[30] Tanaka, T., et al., "2.1 Tbit/s Transmission over 7,221 km with 80-km Repeater Spacing," *ECOC'2000 Proc.*, Post-deadline papers 1.8, Munich, Sept. 4, 2000.

[31] Kobayashi, Y., et al., "A Comparison Among Pure-RZ, CS-Rz and SSB-RZ Format, in 1 Tbit/s (50 × 20 Gbit/s, 0.4 nm spacing) WDM Transmission over 4,000 km," *ECOC'2000 Proc.*, Post-deadline papers 1.7, Munich, 4 Sept. 2000.

[32] Cai, J. -X., et al., "1.12 Tb/s Transmission over Trans-Atlantic Distance (620km) Using Fifty-Six 20 Gb/s Channels," *ECOC'2000 Proc.*, Post-deadline papers 1.6, Munich, Sept. 4, 2000.

[33] Tsuritani, T., et al., "35 GHz-spaced-20 Gbps × 100 WDM RZ Transmission over 2,700 km Using SMF-Based Dispersion Flattened Fiber Span," *ECOC'2000 Proc.*, Post-deadline papers 1.5, Munich, Sept. 4, 2000.

[34] Zhu, Y.. et al., "1.28 Tbit/s (32 × 40 Gbit/s) Transmission over 100 km With Only 6 Spans," *ECOC'2000 Proc.*, Post-deadline papers 1.4, Munich, Sept. 4, 2000.

[35] Ito, T., et al., "6.4 Tbit/s (160 × 40 Gbit/s) WDM Transmission Experiment with 0.8 bit/s/Hz Spectral Efficiency," *ECOC'2000 Proc.*, Post-deadline papers 1.1, Munich, Sept. 4, 2000.

[36] Bigo, S., et al., "5.12 Tbit/s (128 × 40 Gbit/s WDM) Transmission over 3 × 100 km of Teralight Fibre," *ECOC'2000 Proc.*, Post-deadline papers 1.2, Munich, Sept. 4, 2000.

[37] Georges, T., and F. Favre, "Solitons and High Spectral Efficiency," *ROSC International Workshop 2000*, Hakone Sengokura, Japan, Oct.10–11, 2000, www.scat.or. jp/rosc.

[38] Nakazawa, M., et al., "Ultra-High Speed Long-Distance TDM and WDM Soliton Transmission Technologies," *IEEE J. Selected Topics in Quantum Elect.*, 2000, pp. 363–393.

[39] Devaney, J., "Corvis Ups the Ante in the Terabit Regime," *FibreSystems Europe*, Institute of Physics Publishing Ltd., Bristol, May 2001, pp. 28–29.

[40] Ito, T., et al., "Pre-Chirped Normal Dispersion Region Transmission for Highly Marginal Dense WDM Transoccanic System," *OFC'98, Technical Digest Opt. Soc. America*, Paper ThV4, San Jose, CA, Feb. 1998.

[41] Zhu, F., K. Ma, and I. Fisman, "Full-Spectrum WDM Provides Ultra-Long-Reach," *Lightwave*, December 2000, pp. 92–93.

[42] OptimiMight, "OMC 1600 Product Suite," *Supercomm'2001*, Booth 705, Atlanta, GA, June 5–7, 2001.

[43] Chen, D., et al., "6800 km SSMF 10 G DWDM Re-Circulating Loop Using Full-Spectrum WDM Technology and Conventional EDFA," March 2001, http:www.Optimight.com/products/paper_6800fullspec.html.

[44] Emura, K., and T. Ono, "Study on Modulation Formats for Multi-Terabit/s DWDM Transmission Systems," *ROSC International Workshop 2000*, Hakone Sengokura, Japan, Oct. 10–11, 2000, www.scat.or.jp/rosc.

[45] Miyamoto, Y., K. Yonenaga, and S. Kuwahara, "Dispersion-Tolerant RZ Signal Transmission Using Baseband Differential Code and Carrier Suppressed Modulation," *ECOC'98 Proc.*, Vol. 1, Madrid, Sept. 20–24, 1998, pp. 351–352.

[46] Curtis, S., "DWDM Fires Up Cable TV Networks," *FibreSystems Europe*, May 2001, pp. 21–22.

9

Conclusion

Thousand centuries completed, the opportunity will not be closed for anyone to contribute again. All future generations will still find something to discover. In any matter the beginning was far from the achievement, how many objects about which our age hasn't the slightest idea.

—Seneque, *Letters to Lucillius*

"Many books and periodicals contain detailed writings on scientific topics and an increasing amount of technological gadgetry impinges on our daily lives" [1]. In this work, we tried to avoid exclusively giving the tricks of the trade. However we could not avoid being strongly focused on only a small part of the larger optical-telecommunications domain. It is sometimes difficult to step back and see the panorama. At this point we might ask: Where do we go from here? In our domain of DWDM, we have seen during the last few years a bandwagon effect of the business community, after much mistrust in the early 1980s. Today, many venture-capital firms have invested billions of dollars in DWDM and optical telecommunication. But there are likely false trails, on which we are going along, that will be abandoned. The problem is not, as sometimes asked, what will replace DWDM in the long-term? Analysis has shown that satellite communications including low-earth-orbit satellite communications are generally more expensive and capacity limited compared to fiber communications for long-distance high-capacity links. New concepts and new devices will be discovered in the ages to come but it is unlikely that the photonic technologies will be abandoned.

In photonic technologies, using as many frequencies as possible is fundamental. This is fundamental not only because it allows more bandwidth at lower cost but also because DWDM routing and switching already offer the best solution for protection and reconfiguration of large networks. Thus, the progress that is being made on tunable lasers will be one of the main driving factors. In the longer term, packet switching in the optical domain may become a key technology. Optical packet switching would be efficiently combined with Internet and other protocols. DWDM is already widely used in the transport networks and in advanced access networks. It migrates progressively to the metropolitan networks. How to do the migration? We need an effective problem-solving approach. Thus the first step is to ask the right questions. What are the problems to be solved? But unfortunately, "a fool can raise more questions than seven wise men can answer" (proverb). No one expects from me neither to be the only fool raising questions, nor to be one of the seven sages. Instead, to answer the first question: How can we keep in pace with the need for a greater and greater volume of information transfer fueled by the development of communication services moving ever forward? I tried in this book to remain plugged in to the ongoing consensus of the optical-telecommunications community that will progressively decide what are the best solutions.

In the meantime the beautiful and useful scientific research being done in the field excites me. Socrates would have said: "Whatever is useful is beautiful … All that is beautiful is difficult." This is not so different from what was written by William Ralph Inge: "The air that blows around science is like the air of mountain tops, cold and thin, but pure and bracing" [2].

References

[1] Sanitt, N., *Science as a Questioning Process*, Bristol, England and Philadelphia, PA: Institute of Physics Publishing, 1996.

[2] Inge, W. R., "*Living Philosophies,*" New York: Simon and Schuster Ed., 1931, pp. 307–317.

List of Acronyms

AOTF acousto-optic tunable filter

ASE amplified spontaneous emission

ATM asynchronous transfer mode

AWG arrayed waveguide grating

BCPN broadband customer premises network

BER bit error rate

BPF bandpass filter

BPON broadband passive optical network

C^3 cleaved-coupled cavity

CATV cable television

CG concave grating

CO central office

CPN customer-premise network

CS-RZ carrier suppressed return-to-zero

CW continuous wave

DBFA dual-band fiber amplifier

DBR distributed-Bragg reflector

DCF dispersion-compensating fiber

DCM dispersion compensation module

DCS digital cross-connect systems

DFB distributed-feedback

DFG difference frequency generation

DOFA distributed optical fiber amplification

DSF dispersion-shifted fiber

DWDM dense wavelength division multiplexing

DXC digital cross-connect

EDFA erbium-doped fiber amplifier

EDWA erbium-doped waveguide amplifier

ESBG electronically switchable Bragg grating

FBG fiber Bragg grating

FDDI fiber distributed data interface

FDM frequency division multiplexing

FEC forward error correction

FLC ferroelectric liquid crystal

FP Fabry Perot

FPA Fabry Perot amplifier

FSK frequency shift keying

FSWDM full spectrum wavelength division multiplexing

FT Fourier transform

FWHM full width half maximum

FWM four-wave mixing

FXC fiber cross-connect

GCSR grating-coupled sampled reflector (laser)

HDTV high definition television

HDWDM high density wavelength division multiplexing

IFWM intrachannel four wave mixing

IOG integrated optics gratings

IP Internet Protocol

ISDN integrated service digital network

ITU International Telecommunication Union

IXPM intrachannel cross-phase modulation

LAN local area network

LEAF large effective area fiber

LED light-emitting diode

LEDSF large effective area dispersion shifted fiber

LPCVD low pressure chemical vapor deposition

LWPF long wavelength pass filter

MAN metropolitan area network

MEMS Microelectro-mechanical system

MOCVD metal-organic chemical vapor deposition

MONET multiwavelength optical networking

MOVPE metal-organic vapor phase epitaxy

MPLS multiprotocol label switching

MPS multiprotocol lambda switching

MQW multiquantum well

MTBF mean time between failure

MZ Mach-Zehnder

MZI Mach-Zehnder interferometer

NIU network interface units

NOLM nonlinear-optical loop mirror

NRZ nonreturn to zero

NZDSF nonzero dispersion shifted fiber

OADM optical add-drop multiplexer

OBS optical burst switching

OC optical carriers

O/E/O optical/electrical/optical

OF optics filter

ONU optical network unit

OSNR optical signal-to-noise ratio

OTDM optical time division multiplexing

OXC optical cross-connector

PDL polarization dispersion loss

PECVD plasma enhanced chemical vapor deposition

PLC planar lightwave circuit

PMD polarization mode dispersion

PPL passive photonic loop

PSCF positive slope compensating fiber

PWR passive wavelength router

QD quantum dot

QW quantum well

RDF reverse dispersion fiber

RIE reactive ion etching

RT remote terminal

RZ return to zero

SBS stimulated Brillouin scattering

SC supercontinuum

SCF slope-compensating fiber

SCM subcarrier multiplexing

SDH synchronous digital hierarchy

SDL simple data link

SLM spatial light modulator

SM single-mode

SMF single-mode fiber

SOA semiconductor optical amplifier

SONET synchronous optical network

SPM self phase modulation

SPU signal processing unit

SRS stimulated Raman scattering

SSG superstructure grating

SWPF short wavelength pass filter

TDM time division multiplexing

TE transverse electric

TM transverse magnetic

TWA travelling wave amplifier

VC virtual container

VCSEL vertical-cavity surface emitting laser

VHDWDM very high density wavelength division multiplexing

WADM wavelength add/drop multiplexer

WAN wide area network

WBXC waveband cross-connect

WIXC wavelength-interchanging cross-connect

WSXC wavelength-selective cross-connect

WDM wavelength division multiplexing

WTDM wavelength time division multiplexing

XGM cross gain modulation

XPM cross phase modulation

ZBLAN fluoride glass composed of ZrF_4, BaF_2, LaF_3, AlF_3 and NaF

About the Author

Jean-Pierre Laude received his engineering degree in optics from the Ecole Supérieure d'Optique in Paris, in 1963, and his doctorate in spectroscopy from the University of Paris, in 1966. He joined Jobin-Yvon SA in 1966, where he started research on gratings, designed new interferometrically controlled grating ruling engines, and contributed to the new holographic grating manufacturing process. From 1974 to 2000, he served as research director of Jobin-Yvon, as the company worked on spectrometric instrumentation, lasers, gratings, and wavelength division multiplexing (WDM). Dr. Laude established himself as a consultant in optics and spectroscopy in 2000.

Dr. Laude taught graduate courses on dense wavelength division multiplexing (DWDM) at the Ecole Nationale Supérieure des Télécommunications and at the University of Paris. Since 2000, he has been involved in the development of DWDM components for HighWave Optical Technologies.

He contributed to the European research programs Research and Development in Advanced Communications (RACE), and Advanced Communications Technologies and Services (ACTS) in the field of DWDM in different research groups. In 1992, he published his first book on WDM. Dr. Laude is also the coauthor of a number of different books on gratings and optical fiber metrology, as well as an optical encyclopedia. He has published more than 70 papers, and approximately 50 patents on gratings, lasers, spectrometric devices, and WDMs.

He received the SPIE Technology Achievement Award, San Diego, California (1997), for his work in the field of gratings and DWDM, and several awards with the team of Jobin-Yvon (Photonics Circle of Excellence Award 1995, Baltimore, Maryland; Lasers and Optronics Technology Award 1996, Baltimore, Maryland) and with the team of HighWave (European Information Society Technology Prize, 2001, Nice, France) for new multiplexers and routers.

Jean-Pierre Laude is a member of SPIE, IEEE, ASME (the Research Committee on Reliability of Photonics Systems), and of the New York Academy of Sciences.

Index

Recent Titles in the Artech House Optoelectronics Library

Brian Culshaw and Alan Rogers, Series Editors

Chemical and Biochemical Sensing with Optical Fibers and Waveguides, Gilbert Boisdé and Alan Harmer

Coherent and Nonlinear Lightwave Communications, Milorad Cvijetic

Coherent Lightwave Communication Systems, Shiro Ryu

DWDM Fundamentals, Components, and Applications, Jean-Pierre Laude

Fiber Bragg Gratings: Fundamentals and Applications in Telecommunications and Sensing, Andrea Othonos and Kyriacos Kalli

Frequency Stabilization of Semiconductor Laser Diodes, Tetsuhiko Ikegami, Shoichi Sudo, and Yoshihisa Sakai

Handbook of Distributed Feedback Laser Diodes, Geert Morthier and Patrick Vankwikelberge

Helmet-Mounted Displays and Sights, Mordekhai Velger

Introduction to Infrared and Electro-Optical Systems, Ronald G. Driggers, Paul Cox, and Timothy Edwards

Introduction to Lightwave Communication Systems, Rajappa Papannareddy

Introduction to Semiconductor Integrated Optics, Hans P. Zappe

LC3D: Liquid Crystal 3-D Director Simulator, Software and Technology Guide, James E. Anderson, Philip E. Watson, and Philip J. Bos

Liquid Crystal Devices: Physics and Applications, Vladimir G. Chigrinov

Optical Document Security, Second Edition, Rudolf L. van Renesse, editor

Optical FDM Network Technologies, Kiyoshi Nosu

Optical Fiber Amplifiers: Materials, Devices, and Applications, Shoichi Sudo, editor

Optical Fiber Communication Systems, Leonid Kazovsky, Sergio Benedetto, and Alan Willner

Optical Fiber Sensors, Volume Three: Components and Subsystems, John Dakin and Brian Culshaw, editors

Optical Fiber Sensors, Volume Four: Applications, Analysis, and Future Trends, John Dakin and Brian Culshaw, editors

Optical Measurement Techniques and Applications, Pramod Rastogi

Optoelectronic Techniques for Microwave and Millimeter-Wave Engineering, William M. Robertson

Reliability and Degradation of III-V Optical Devices, Osamu Ueda

Smart Structures and Materials, Brian Culshaw

Surveillance and Reconnaissance Imaging Systems: Modeling and Performance Prediction, Jon C. Leachtenauer and Ronald G. Driggers

Understanding Optical Fiber Communications, Alan Rogers

Wavelength Division Multiple Access Optical Networks, Andrea Borella, Giovanni Cancellieri, and Franco Chiaraluce

Printed in the United States
77481LV00002B/238

9 781580 531771